The Blue Book

Stamatina Th. Rassia
Editor

The Blue Book

Smart sustainable coastal cities and blue growth strategies for marine and maritime environments

Foreword by Nick V. Baker

Springer

Editor
Stamatina Th. Rassia
Le.D.R.A. Group
Athens, Greece

ISBN 978-3-031-48830-6 ISBN 978-3-031-48831-3 (eBook)
https://doi.org/10.1007/978-3-031-48831-3

This Springer imprint is published by the registered company Springer Nature Switzerland AG
The registered company address is: Gewerbestrasse 11, 6330 Cham, Switzerland

Paper in this product is recyclable.

Foreword

The topics of this book fit nicely into a three-circle Venn diagram – the ocean, the coast and the ship. Setting the widest geographical context, shipping in the ocean ecology is examined, and impacts ranging from the carbon emissions of the global shipping fleet, to the effect of ships' propeller noise on whale communications, are described. The authors clearly anticipate that shipping will continue to be the main method of transport and indeed anticipate growth. However, although marine carbon emission efficiencies may improve due to technical developments, including electrification, we must not forget that this demand for sea transport is largely fuelled by the consumption of goods whose production in the east is responsible for a large proportion of global manufacturing emissions.

The description of the ecological state of the oceans makes some depressing reading: overfishing, habitat loss, invasive species and microplastics. There are some hopeful signs, however; for example, sophisticated ship management schemes using a carbon intensity index to indicate emissions, controls on ship's antifouling paint and water ballast management to prevent the introduction of invasive species. The economic case for concern is made – fish stocks, tourism, as usual. But I was pleased to see, alongside the rational arguments, appeals made to our sense of the beauty and value of the sea and its wild inhabitants and the need for public engagement, education and enlightenment.

Moving to the coast, one is struck by how in the minds of the techno-oriented city planners even the simple "day by the seaside" needs the application of smart systems to control everything from catching the bus to whether the waves are ok for surfing! With the exception of the later example, I found a few other issues that where uniquely related to coastal cities. Similarly, overheating mitigation by controlling the albedo of building surfaces clearly applies to all urban developments. But the discussion of how predictable diurnal sea breezes impact the urban heat island and could be taken into account for both indoor and outdoor ventilation design, is highly relevant and has a solid historical precedent in coastal towns.

Prompted by the recent COVID-19 pandemic (graphically illustrated by the description of the cruise ship Diamond Princess with 3500 passengers and crew on board quarantined for 2 weeks off the Japanese coast), the emphasis of this book is undoubtedly on the health of the passengers, in particular on cruise ships. The impact on the cruise industry has been massive, and this has prompted intense research interest at a European level, as typified by the Healthy Ship for You (HS4U) EU-funded project.

To me, as a building scientist used to being concerned with the health and well-being of the occupants of buildings, the thought of several thousand people in close proximity in a deep, multi-story steel "building" with serious constraints of the free access to light and air demanding fully mechanical ventilation sounds like a recipe for disaster. Throw in the fact that they all eat food from a common kitchen, meet for eating and entertainment, and then are released in their thousands, every few days to go ashore and interact with another population, which would seem to ensure disaster.

Several chapters deal with this, and in all cases, solutions are sought in the detection, management and mitigation by engineering and IT systems, rather than ship design itself. This is not surprising since unless cruise ships became much smaller, many of the options open to the land-based architect, relating to passive access to natural light and air, are simply not available.

As expected, IT plays an ever-increasing role in proposed solutions. Disease detectors are discussed at length, and it may come as a surprise to the uninitiated to learn that the risk of a passenger having COVID can be sensed by analysing an acoustic record made in their cabin, of their sneezing and coughing habits. Other sensors even invade the toilet, but it is reassuring to know that it will be "listened to" by AI, and not the medical officer! IT does no stop there. The next step described is the application of the data to a mathematical cross-infection model of the ship to predict risk and then test various management interventions (such as isolating a cabin or group of cabins). Finally, in the most ambitious proposals, the system could carry out some responses (e.g. changing ventilation strategy, closing down areas, activating warnings, etc.), automatically.

Whilst still on board, one further topical issue deals with is the use of (possibly redundant) ships for permanent or semi-permanent accommodation, as an alternative to housing. This mainly focuses on the use to house asylum seekers; clearly, the economics of this is skewed by the strong political propaganda pressures. Where this is absent, moving to the normal housing market, it seems that there is no great demand for ship-borne accommodation other than at the super-luxury end of the spectrum.

It is interesting to note that many of the problems of cross-infection to be found on cruise ships could be present in residential ships – particularly in the case of refugee/asylum-seeker accommodation, which is normally at very high density. It is also worthy of note that in the UK, the purpose-built asylum-seekers' barge, the Bibby Stockholm, moored in Portland Harbour, UK, to house 500 people, was closed down within a few days of initial occupation on various grounds, including fire safety and infection risk.

On this topic of health – a final irony. The historic precedent for this amount of interest into the health of ships "passengers", was in the 18th century – the provision of ventilation to slave ships – not out of concern for the unfortunate slaves but rather as a way of ensuring the valuable "cargo" survived and was in good condition for resale.

So, after reading the book, am I in a position to speculate on the Blue Future? It is worrying to me that the anticipated future of shipping assumes the ever-increasing growth of trade and ship movements. The presence of the human species on land is already far outside any ecological limits and is already threatening the oceans. I would prefer a far gentler future, with a gradual diminution of the transport of inessential low-value goods, or carbon-emitting oil, across the globe. On the other hand, the return of passengers to surface travel (for which the strength of the cruise industry indicates a growing demand) could mean that sea voyaging once more becomes an enriching experience for many and an opportunity for them to confront a true wilderness.

Continuing my dream, I think ships will get smaller again, maybe mixing essential cargo-carrying with passengers, and the reduced size will allow much simpler approaches to environmental conditioning (more akin to passive design to be found in the latest low-energy building). And two resources that are freely available at sea are wind and sun, so a combination of sails, photovoltaics and turbines with battery storage would power the ships at energy-efficient speeds with zero carbon emissions. Smaller ships have other advantages – requiring smaller port facilities and being closer to the social centre of the town – but I'll still allow a smart transport system to pick up passengers.

The authors of this book have homed in on a range of essential and interconnected issues related to our use of the oceans. It is reassuring that all point to an environmental objective, be it a healthy individual or a healthy planet.

Architectural Association
London, U.K.
6 October 2023

Dr. Nick V. Baker

Preface

The highest percentage of the Earth's surface is covered in water (about 70%), of which about 97% is ocean water. Over-engineering and exploitation have led to the development of about 40% of the world's cities near coastlines. While the oceans act as the coolers of the atmosphere, global warming and the vast melting of ice sheets will soon bring coastal regions to the forefront of a series of crises. Oceans have absorbed more than 90% of the excess heat in the atmosphere. Ocean warming is now on the cards. Heating up the marine ecosystems can lead to food scarcity, poverty and finally global insecurity. While scientists have stressed the dangers ahead of us, habitats continue to grow and to deteriorate land and water ecosystems. Global societies seem at the same time to remain blissfully unaware of what is yet to come. Public health crises add to the ever-growing picture that threatens contemporary lifestyles.

"The Blue Book: Smart Sustainable Coastal Cities and Blue Growth Strategies for Marine and Maritime Environments" aims to address global challenges of sustainability and health in cities, coastal areas, marine ecosystems and maritime environments. It presents a unique combination of new ideas, expert knowledge and research results on a variety of related fields, such as naval architecture, marine engineering, biomedical informatics, public health, architecture engineering and building physics, nanotechnology as well as ecology and innovation. It is composed of 12 chapters written by eminent researchers and practitioners from Australia, China, Israel, Bangladesh, Luxemburg, Germany, France, Portugal, Malta, Italy, Cyprus and Greece. Some chapters include research results, which have been produced recently within the new cutting-edge EU programme called "Healthy Ship 4U" (GA: 101069937).

Overall, this book offers a wide spectrum of interdisciplinary topics that appear under the following titles, by the respective authors and author groups.

- *Shipping and Ecology* by Gregory J. Grigoropoulos
- *Ocean Wildlife and Megafauna Protection* by Mohammad Zahirul Islam
- *Big Data and Sustainable Applications for Healthy and Safe Coastal Cities* by Amalia Ntemou, Akrivi Kiousi and Nasos Grigoropoulos
- *Developing Housing at Sea: A Case for Humanitarian Assistance and Residency Vessels* by Kaithlyn West and Rita Cheung
- *Smart Coastal City Mobility Management in the Context of High Tourist Activity and Methodological Approaches* by Bertrand David, Chuantao Yin and René Chalon
- *Urban Climate and Mitigation in Coastal Cities* by Mattheos Santamouris and K. Vasilakopoulou
- *Sick Ships: A Discussion on Historical Cases and Optimization for the Future* by Stamatina Th. Rassia and Thodoris Emm Tsikis
- *Disease Spread Control in Cruise Ships: Monitoring, Simulation, and Decision Making* by Georgios Triantafyllou, Panagiotis G. Kalozoumis, Eirini Cholopoulou and Dimitris K. Iakovidis
- *Real-Time, Systematic Disease Detection on Cruise Ships: Feasibility Assessment for Outbreak Prevention* by Bérengère Lebental
- *Technology Approaches for Cruise Ship Disease Propagation Monitoring* by Bruno Almeida, Márcio Mateus, Miguel Pacheco, Pedro Maló and Tiago Teixeira
- *Scenarios for Sensors and Actuators Deployment to Prevent and Mitigate Epidemics on Cruise Ships* by Bérengère Lebental, Kaitlyn West MPH, Olga Vainer, Michel Hoffman, Stathes Hadjiefthimiadis, Babis Andreou, Marc Bonazountas, Iro Palaiokosta, Astrinos Papadakis, Nasos Grigoropoulos, Amalia Ntemou, Anne-Sophie Fölster, Dimitris Drikakis, Bruno Almeida, Márcio Mateus, Miguel Pacheco, Pedro Maló and Tiago Teixeira
- *Improving Biological Safety on Ships Using Nanotechnology* by Arturo Sommeriva, Bruno Cantarelli and Costantino Gagliardi

By "painting" the cover of this book blue, the editor wishes to manifest the ever-growing value of looking into the marine and maritime ecosystems aiming to tackle long-standing challenges of sustainability and multilevel developmental needs.

Finally, we would like to cordially thank all chapter authors for their collaboration and expert contributions. Last but not least, we would like to express our sincerely gratitude to the superb assistance of the Springer staff during the preparation of this publication.

Athens, Greece Dr. Stamatina Th. Rassia

Contents

Shipping and Ecology (Sustainability)

Gregory J. Grigoropoulos

1 Introduction

More than 2500 years ago, waterborne transportation constituted the dominant mode of trading, at least in the Mediterranean Sea region, which encompassed many highly populated cities along the shoreline. At that time, the most powerful Greek cities established colonies with which they kept close political, economic, cultural, and trading relations. Later, European trade extended into Asia and North Africa based on land routes, travel on which, however, was time consuming and restricted with respect to the bulk of the goods transported, and thereby expensive.

During the Renaissance, new commercial sea routes were established across the world and trade grew enormously. During the nineteenth century, the replacement of wind as the main propulsion source by steam, and later by the internal combustion engine, allowed for the production of faster and more reliable merchant ships. In addition, the introduction of steel as the principal construction material instead of wood enabled the building of large ships capable of transporting huge quantities of cargos.

Nowadays, waterborne transportation remains the prime mode of international trading. Thus, merchant ships transport 90% of the goods as well as a significant percentage of global passengers. Even in Europe, 30% of intra-European trade is carried out by sea [2].

Adam Smith propounded the economic significance of the shipping sector, claiming that it offers a cheap mode of transportation, which in turn helps the access of larger markets. Indeed, the significance of shipping has not deteriorated despite the relatively long trip times of sailing compared with other modes because the maritime community has enough capacity to feed the markets in a smooth and

G. J. Grigoropoulos (✉)
National Technical University of Athens, Athens, Greece
e-mail: Gregory@central.ntua.gr

S. Th. Rassia (ed.), *The Blue Book*, https://doi.org/10.1007/978-3-031-48831-3_1

sufficient way. The huge quantities of goods and large numbers of passengers carried by contemporary ships constitute the other major advantage of sea transport over its land-based and airborne counterparts.

Although waterborne keeps on being the prime transportation mode, the combination of waterborne with airborne and land-based transportation means (trucks, trains) has begun to create new opportunities for efficient intermodal transportation; however, this necessitates improvements in airport, road, and rail infrastructure, leading to further development in the countries. In light of the foregoing, the perspectives and the future of shipping seem to be promising, although they still depend heavily on the vicissitudes of international trade, the evolution of political affairs, and the development of domestic economies in conjunction with the available tonnage capacity.

The strength of shipping as a major commercial activity and its potential for future growth, are widely accepted; to this target contribute also the milder environmental effects of sea routes compared with rival transportation modes such as land and air. Indeed, considering the intense interest in environmental pollution, merchant and passenger ships pollute less than other modes of transportation. At present, shipping is responsible for less than 3% (around 2.6%) of air pollution via the carbon (CO_2, CO, C) and Green House Gases (GHG) emissions, being a small portion, of the overall emissions.

2 Strategies to Emission Reduction

The operation of ships has been subject to strong antipollution legislation since the establishment of the MARPOL International Convention in 1978. However, shipping was embodied in the emergency measures initiated by the Organization of United Nations (UN) to protect the environment and to reverse the intense trend for destabilization of the world climate due to an unusual increase of the average temperature by 2 °C during the current century. These measures, consider as reference the situation at the beginning of the century and try to exploit current practices, as well as high-tech and innovative technologies to counteract to this very disappointing perspective.

The International Maritime Organization (IMO) initiated an effort to drastically reduce emissions due to shipping, despite the expected increase in the bulk of waterborne transport. To this end, IMO imposed additional regulations for continuous monitoring and reduction of the fossil fuel consumption of ships which can be fulfilled by:

- sailing at reduced speed (slow steaming),
- using additional appendages to improve hull and/or propulsion efficiency,
- exploiting wind (sails, Flettner rotors, kites) and sun (Photo-Voltaic panel) as auxiliary energy sources,

- collecting the CO_2 and the GHG emissions in a form to be delivered to a disposal station,
- any combination of the above measures.

All these measures, except for the collection of the emitted gases, are directly connected to an analogous decrease in fuel consumption, which will lead to more economically efficient operation of ships.

Furthermore, 15 years ago, IMO introduced as targets the reduction of the CO2 emissions by 40% till 2030 and by 70% till 2050, with reference the year 2008. The respective GHG emissions should be reduced by 50% till 2050.

To provide a regulatory framework to achieve these targets, existing (and proposed) amendments to the Pollution Prevention Treaty (MARPOL) took a technical and operational approach to reducing GHGs. These include:

- Energy Efficiency Design Index (EEDI): New ships must be built and designed to be more energy efficient.
- Ship Energy Efficiency Management Plan (SEEMP): A practical tool for helping shipowners manage their environmental performance and improve operational efficiency.
- Energy Efficiency Existing Ship Index (EEXI): Set to enter into force in 2023, EEXI applies many of the same design requirements as the EEDI, with some adaptations regarding limited access to design data.
- The Fuel Oil Consumption Data Collection System (DCS): Mandates annual reporting of CO_2 emissions and other activity data and ship particulars for all ships above 5000 GT. The annual reporting of CO_2 emissions started on the 1 January 2023.

Additional measures include the characterization of some sea areas as Sulphur Emission Control Areas (SECA), with allowable sulfur emission 0.10% m/m, compared to 0.50% m/m in the rest sea areas.

More recently, IMO Marine Environment Protection Committee (MEPC) in its 78th session introduced the Carbon Intensity Indicator (CII), a rating scheme (A-E) developed by the IMO to measure the annual performance of all ships above 5000 GT in terms of CO_2 per DWT and distance covered. This scheme intends to directly correlate sustainability to transport capacity of shipping.

Meanwhile, in 2022 the European Union (EU) issued the “Fit for 55” package, as part of the European Green Deal, aiming at reducing the EU’s GHG emissions by 55% till 2030. In addition, EU requested that ships above 5000 GT, regardless of flag, arriving at, within or departing from EU ports and/or European Economic Area ports should submit the MRV Regulation forms for the monitoring, reporting, and verification of CO_2 emissions. Although the MRV Regulation and the IMO data collection scheme (DCS) established under Regulation 22A of MARPOL VI are both intended to capture fuel use data in order to quantify carbon emissions from shipping they are significantly different, with the unfortunate result that ship operators will have to manage two separate reporting schemes for the fuel they use. Further guidance on the use of MRV in shipping is provided by the EC who established the European Sustainable Shipping Forum (ESSF) subgroup for this purpose.

3 New Technologies in Shipping Serving Sustainability

Shipping community expect that current trends and associated emerging technologies will contribute significantly to the sustainability of waterborne transportation. In the sequel, we present in short these new technologies and then describe their expected impacts on society and shipping. We followed a two-step approach in identifying these trends and technologies. Our first step (top-down) looked at the land industries and those that are adjacent and considered technology areas and trends likely to be influential to shipping. Our second step (bottom-up) drew on the insights of the scientific community to identify technologies that will have an impact before 2025 and those that we see emerging toward 2030. The opinion of the stakeholders is taken into account to validate our findings.

The selected technologies are grouped into seven key technology trends and further into three categories:

ENABLING: Three technology trends can be thought of as "enabling" and fostering other technologies: acceleration of digitalization, virtualization and automation across the life-cycle, and precision materials.

TRANSFORMING: Two technology trends can be thought of as transformative in key industries: transport and logistics, low carbon energy systems.

SUSTAINING: The third group of technology trends contains technologies that help to sustain our biosphere: the ocean space and other natural ecosystems.

For each trend, we have selected those technologies that have the largest expected economic and societal impact on our target industries (Maritime, Oil and Gas and Energy) in the coming decade.

We assume that the world will follow current trends toward 2030, with political and environmental uncertainties shaping the global risk picture. Technologies typically introduced to create advantages, also add new risks [3]. Key questions are therefore:

- *Which technologies will emerge that create new risks and uncertainties affecting the long-term value of our assets as well as the well-being of people and the planet?*
- *And, which emerging technologies could help us manage our risks better and help us to reach the SDGs?*

3.1 Acceleration of Digitalization

Today, we can detect the signature of six digital technology trends that will shape our world toward 2030 and 2050 [1]:

- Connecting people to the internet
- More intimately ubiquitous sensors (IoT):
- Communication and computing at every place
- Artificial intelligence
- Platform-based social and economic models
- The digitization of material through additive manufacturing

These technologies are accelerating and enabling the development of other technologies at a pace never seen before. As we create and innovate at ever greater speed, are we losing the ability to reflect on the impact of our creations in a broader societal context?

For all the opportunity it brings, the acceleration of technology through digitalization generates economic, social, and ethical challenges. Before society has had a chance to reflect on the broader implications of a given technology, it is already packaged for sale in a variety of channels. Today, anyone can get free access to a cloud provider and begin experimenting with very powerful machine learning capabilities, without thinking about things such as gender or race bias. And, while enterprises struggle to manage the explosion of technologies coming to market, new business models are disrupting existing businesses more rapidly than ever.

Compositional architecture is the key concept to the success of enterprises adopting emergent technologies. It refers to the ability to take proven and emerging technologies, combine them with data, and develop new business models to fulfil raising expectations of the market. Incorporation of successful features to the design and operation of ships will be improve life value of ships. The cooperation with the ports involved in their routes will further improve the logistics of shipping.

The pipeline of data, from creation to value extraction, will become an even greater opportunity for organizations managing their data ecosystems, as well as for companies creating technologies for each part of the value chain. The data ecosystem has three distinct parts:

Connectivity consists of the IoT combined with the networks the data traverses. With 5G networks, enterprises will be instrumented and measured.

Processing and management are fundamental to every organization, encompassing the data they create, consume, and share. Tools providing data quality management, provenance, and compliance to organizational and regulatory standards will form a key part of any enterprise's architecture.

Presentation and use of the value extracted from the data pipeline. Examples are dashboards, mixed reality remote assists, and automation of normally manual operations in factories and plants.

3.2 Virtualization and Automation Across the Life Cycle

Virtualization and automation are being taken to entirely new levels, creating the basis for major transformation across all industries and our wider society. High-level abstraction, enabled by digitalization, allows us to connect a wide variety of virtual items to create increasingly connected and efficient processes.

Virtual presence and tele-presence will change many work routines before 2030 – not least the inspections, audits, and assessments carried out by certification bodies. It will be possible to perform such activities both remotely and with greater "view" and "depth," due to advanced camera systems and sensors combined with AI to aid decision-making. In a wider industry context, meetings, conferencing, and training will have moved firmly into the virtual reality space by 2030, aided by a range of wearable technology, and spurred on by the need to cut travel-related expenses and emissions.

Developments within Industry 4.0 (including advanced sensor technology and machine learning algorithms and strengthened inclusion of feedback loops) will see the remote and autonomous operation of most of systems intensify.

Automation, autonomy, and remote operation will coevolve, with the related technical disciplines challenging and fertilizing each other.

Toward 2030, virtualization will connect all elements and dimensions of the system life cycle, aided by digital twins. These will function as a collection of all available system data and the models used for decision support throughout the life cycle and made available to dedicated stakeholders.

The inventory of enabling technologies in this sector is staggering, and includes:

- Alternative low- or zero-carbon fuels
- Decentralized production of parts and structures through additive manufacturing
- Artificial Intelligence (AI) and its subset – machine learning
- Augmented, virtual, and mixed reality
- Blockchain technology that can revolutionize contracts, tracking and payments
- The Internet of Things (IoT)
- Quantum computing for logistics optimization
- Robotics assisting greater automation, and
- More powerful, reliable, and quicker digital communications and geolocation with advanced satellites

3.3 Toward Precision Materials

Materials science, and its applications in novel and improved technologies aims to provide abundant, affordable, and environmentally friendly essentials for life. Materials for tomorrow will need to meet the demands of fabrication while remaining cost efficient and complying with sustainability and safety requirements. The cost of newbuilding, raw materials shortages, and requirements to limit environmental footprints, will all drive technology that enhances life extension and reuse, and which promotes recycling of assets to a much greater extent than seen today. Recycling is critical in the circular economy model, where waste is minimized through maintaining the persistence of technical nutrients (i.e., metals, plastics, ceramics, etc.). However, an ever-greater emphasis is being placed on designing products for longevity, shareability, repair and servicing, and remanufacturing. A core principle of circularity is that fewer materials are used, and less energy expended if a product can be easily repaired rather than discarded (where materials first must be extracted before being incorporated new products).

One glaring example of the failure of the linear consumer economy to responsibly manage materials is the increasing problem of plastic waste in our oceans, where remedial actions are now needed, but are logistically and economically complex and demanding.

3.4 Toward Future Transport and Logistics

The decade to 2030 will be remembered as a major turning point in the history of mobility and logistics. The technologies driving this shift are intimately bound with new scientific advances, ways of thinking, and business and economic models arising from the ever-more urgent mission to "save the planet."

Barely a week passes without the major headlines about developments in electric vehicles (EVs), with sales in parts of the world achieving new records. The first battery-powered passenger aircraft is already in the air, with commercial operation closer than many realize. Transport as a service, paid for per-use rather than owned, is a natural consequence of the connectivity boom.

Electrification is a major trend in the transportation sector. It is currently most prevalent in private vehicles in areas with a mature battery recharging infrastructure.

Passenger EVs are on course to become mainstream, at least in Europe, by 2024, and the enormous two- and three-wheeler fleet in India, China, and south-east Asia will be 90% electric by 2030. With a cost learning rate (i.e., the rate at which costs fall with every doubling of capacity) of 19%, costs for batteries will plunge faster than even those for solar PV and wind technology, especially with the advent of solid-state chemistries before 2030.

Further improvements in the cost, energy density, weight, and volume of electric batteries will enable wider use of battery-storage systems in heavy road transport such as trucks and buses, and in short-haul air travel and short-sea shipping.

On the face of it, the transport revolution seems entirely positive: cheaper, connected mobility, and logistics that offer a wide range of benefits and choice to businesses and consumers – while making major strides toward the decarbonization of economic activity. However, the fundamental role of transport and logistics in society makes the sector especially vulnerable to cyberattacks, and there is a mounting need to identify potential risks and implement appropriate safety measures. There are also unresolved ethical dilemmas associated with the design, operation, and maintenance of self-driving vehicles.

Many of the disruptions in mobility are driven by big tech, representing a further avenue for concentration of power, data, and control for entities that are considered by some to be digital "monopolies."

3.5 Toward Low Carbon Energy Systems

Fossil fuels have dominated the world's primary energy supply for the last century, but a transition to a more integrated and low carbon energy system is now gathering momentum. The transition may seem slow because the replacement energy providers (mainly wind and solar PV) start from a very low share. Indeed, by 2030, fossil sources will still comprise more than 70% of the energy mix, and continued investment in new sources will be critical, as will the quest to lower the carbon footprint of fossil fuel production and consumption.

However, the pace of renewable energy development is accelerating, both technologically and in terms of affordability, such that the next decade will witness some key tipping points in the energy landscape.

The present shift in the energy mix is both market and policy driven, with climate-related concerns, amplified by youth activism, expected to have a significant regulatory impact. Local pollution and environmental problems play also a role in driving this transition, in addition to a geopolitical trend of reducing dependency on energy imports. However, the contribution of shipping to local emission is minimal, since the main sea routes are located far away from populous regions, which mainly suffer from environmental pollution and are more sensitive to such problems.

The shift toward low carbon energy systems is taking place in all industry sectors and across all world regions. Many of the drivers and technologies are well known, but there are nascent technologies, with more uncertain futures that could emerge and scale over the coming decade. These could significantly reinforce or delay the transition away from fossil fuels, depending on their rate of development.

In the period between now and 2030 almost all (90%) net energy additions are expected to come from renewable energy sources, mainly wind and solar PV. These additions will start to affect the mix and compete with existing energy carriers. Furthermore, energy demand has reduced by 1.6% annually over the last two decades and is expected to accelerate toward an annual reduction of 2.5% by 2030. This is principally driven by the growing renewable share in electricity generation, eliminating enormous heat losses, and electrification of the energy system generally.

3.6 *Transition Drivers*

Policy A growing spectrum of government policies and targets, implemented through incentives, taxes and specific regulations. Nationally Determined Contributions (NDC) and the Paris Agreement are likely to be strengthened from 2020. Other policies linked to local pollution, energy security, and independence will increasingly focus on the role of renewables in satisfying these aims.

Investment Community Intensifying focus on the implications of oil demand peaking, and the risk of stranded assets, coupled with growing demands for climate-related financial disclosure. The investment community will increasingly require companies to stress test their business models based on different low-emission energy futures.

Technology and Costs Rapid developments such as advancing digitization and connectivity are enabling reduced costs and contributing to the speed of this transition. Renewable energy has seen dramatic cost reduction the last 10 years, and further significant reductions of solar or wind generated electricity cost are expected. Wind is increasingly going offshore, initially with fixed installations like today, but eventually floating wind farms will become cost-competitive, in addition to not contending with other land use.

The extractive fossil fuels industry itself is currently transforming to respond to the technological and policy changes. Coal-fired generation investments have plummeted 75% since 2015 and many companies in oil and gas are transitioning toward the less-polluting gas value chain. Increasing gas demand leads to gas transport on keel more than doubling toward 2030, leading to stabilization in gas prices globally.

Decarbonization With oil and gas still dominant in 2030, there will be increased focus on decarbonization in areas where emissions reduction is difficult to achieve by fuel switching, and thus there will be growing demands to further develop carbon capture and storage (CCS), where CO_2 is captured and stored safely to obtain a net removal from the atmosphere, using a range of technologies. Negative emissions are needed to handle the overshoot of atmospheric CO_2, provided that all mitigation options are pursued, to avoid further global warming-induced climate catastrophes.

3.7 *Transition Barriers*

Lock-in and inertia of the existing energy infrastructure, where existing fossil fuel subsidies and a need to protect both energy and job security are significant barriers to change.

A Pivot from Opex to Capex Renewables require high up-front investments, but those are offset by lower operational costs compared with fossil fuels. The extra risks associated with any investments in an environment undergoing rapid technical and policy change can make it challenging to access capital, especially in developing economies.

Power Market A higher mix of renewables tends to lower electricity prices in certain periods, but variable resources may then struggle to secure profits from further investments without changing the structure of the power markets. This is amplified by internal competition between these sources, such as low-cost wind competing with even cheaper solar PV.

3.8 Toward Wider Use of Ocean Space

The oceans cover around 71% of the Earth's surface and are an important source of food, minerals, and energy. However, several factors threaten the ocean ecosystem, including overfishing, pollution, and acidification.

Ocean health is under severe pressure. A better understanding of marine ecosystems is crucial to achieving more sustainable ocean management. With only 5% of the oceans mapped in detail, new technologies will facilitate the mapping and extraction of ocean-based resources. Autonomous systems are ideally suited to explore and monitor ocean resources in areas posing logistical challenges.

As ocean use intensifies and new ocean areas open up, new industry sectors are moving offshore, and new structures are being developed to accommodate them. The oceans also represent an area of development for crowded coastal areas. Floating cities may offer a housing solution in response to the effects of climate change.

3.9 Ocean Health Under Severe Threat

According to a report by the Intergovernmental Science-Policy Platform on Biodiversity and Ecosystem Services, plastic pollution has increased tenfold since 1980, and 300–400 million tonnes of heavy metals, solvents, toxic sludge, and other wastes from industrial facilities are dumped annually into the world's waters. Fertilizers entering coastal ecosystems have created more than 400 ocean "dead zones," totaling more than 245,000 km^2 (a combined area greater than that of the UK). A transformation to sustainable practices is crucial to preserving productive and resilient oceans and seas. The report "Global Goals, Ocean Opportunities" from the UN Global Compact elaborates on how to protect and preserve the ocean while achieving increased production [4].

Toward 2030, the OECD estimates that the global value added in the ocean economy will maintain its share of world total GVA at around 2.5%. These estimates are conservative, and do not include some of the ocean-related industries for which adequate data are not available. For the same reason, the estimates understate activity in certain industries (such as shipping) for which numerous countries have had to be excluded. The modest growth or even contraction expected in some large industries (e.g., offshore oil and gas) masks comparatively high rates of growth expected in others (e.g., marine aquaculture, offshore wind, fish processing, port activities) and holds back overall average growth in the ocean economy. Many segments in the ocean economy have the potential to outperform the growth rate of the global economy.

In 1982, Greek shipowners established a nongovernmental organization in order to promote shipping quality and environmental protection and protect the marine environment. This organization, known as the Hellenic Marine Environment Protection Association (HELMEPA), has been active both within Greece and in international institutions such as the IMO. Similar Associations have been founded in other coastal countries, and in 2006 the International Marine Environment Protection Association (INTERMEPA).

The European Parliament and Council issued the EU Directive 2008/56/EC for the protection and management of the sea environment. Further, initiatives on the port side have led to increased environmental protection and management options. For example, in collaboration with the European Sea Ports Organization (ESPO) and the EcoPorts Foundation, the major ports of Europe have been certified by the Port Environmental Review System (PERS) accreditation system for their environmental management systems. PERS doesn't only incorporate the main general requirements of recognized environmental management standards (e.g., ISO 14001), but also takes into account the specificities of ports.

Last but not least is the interest of IMO for minimizing the effect of ship operation on the health of the oceans. To this end, IMO issued guidelines for the control and management of ships' biofouling to minimize the transfer of invasive aquatic species. To this end IMO imposed the installation of Water Ballast Management Systems for the treatment of the ballast water so that the ballasting of a ship at a sea area and its deballasting at another one doesn't result in the transfer of pathogenic organisms from the former to the latter sea area. Finally, anti-fouling coatings on the underwater hull surface that are dangerous for fishery and other organisms living in the seas.

3.10 Impact on Maritime

In the next decade, seaborne trade demand is predicted to grow by 26%, to reach almost 80 Gt-nm by 2030. This steady growth will coincide with a decade of significant steps toward decarbonization and investment in digitalization. All maritime stakeholders will be challenged to significantly reduce GHG emissions, forcing strategic choices, many of which will need to be made well before 2030, on low- to zero-carbon fuels and technologies to meet the IMO goals for 2050. By 2030, digitalization will have impacted all activities in the industry – but not equally.

Waterborne cargo transport will be increasingly integrated into digital supply chains; whereas for other waterborne segments, like offshore services and cruise, reducing the challenge (and opportunity) is to master the digital complexity of assets. Technologies that stand out as game changers are:

- Fuels and associated technologies
- To reduce carbon emissions
- Additive manufacturing
- Digitized shipping, and
- Safety of cyber-physical systems

It must be stressed, however, that attaining high quality in the shipping industry is an extremely difficult and painstaking process since it involves international cooperation, and it is affected by many factors. Major maritime nations, in particular, face difficulties due to globalization, while intense and often fierce competition in freight trends compete with quality shipping services. The numerous legal chances open to shipping firms, which allow them to expand their activities into different regions of the world, lead to anonymity and increased difficulties when investigating pollution incidents or the breaching of contractual duties.

Further, the ease with which a shipping firm can change registries (e.g., by selecting those that allow increased profits while minimizing running and maintenance costs) seems to be one of the most serious problems facing the industry today. Many OECD countries, including Greece, operate open registries. In addition, many EU countries resort to international as well as parallel registries in order to minimize the losses incurred by their national registries. This situation not only affects the stability and improvement of the maritime sector but also poses a serious threat to maritime safety and the marine environment.

Quality shipping in the modern maritime industry seems to be a difficult objective achievable only through international cooperation between OECD and EU countries. Therefore, it seems reasonable to suggest that the only way forward is to strive for quality shipping throughout the world. This would mean avoiding the split that exists in the maritime sector between first class shipping (i.e., serving OECD and EU countries) and second class, comprising lower-standard ships under flags of convenience, mainly owned by firms originating from EU and OECD states that serve the rest of the world.

3.11 Impact on Oil and Gas

Technology developments and impacts for the oil and gas industry to 2030 will be driven both by economic factors and by sustainability issues. The industry faces an unprecedented combination of market forces, regulation, societal pressure over climate change, and challenges to integrate with other sectors like power and renewables.

The following technologies, among others, can help the sector to meet three key sustainability challenges: efficiency, decarbonization, and safety:

- The emergence of a digital oil and gas value chain run by machines and software algorithms
- Wider adoption of low-carbon energy carriers, such as hydrogen for heating and transport
- Subsea facilities becoming all-electric, with huge capital cost savings for field developments

3.12 Discussion and Conclusions

Shipping is a powerful and highly competitive economy sector, which actually serves the world commerce. Although its power is growing steadily, the respect to the environment turned recently from a secondary necessity to a prime focusing point.

Among the various environmental consequences of ship operation, the most demanding is the emission of carbon oxides and GHG. The use of alternative non-fossil fuels, without carbon content to propel ships and the carbon capture technologies are quite promising in this respect. The use of electric motors as prime movers in conjunction with a smart central control of the overall power needs onboard using digitalization and automation seems to be corner stone of any environmentally friendly ship operation. The production of energy by Internal Combustion Engines (ICE), air turbines, or fuel cells (FC), as well as its storage in batteries, depends on the evolution of the respective technologies during the ongoing and, may be, the following decade. The winning candidate technology will be first conquering the power stations at land and then it will be marinized for implementation at sea.

References

1. DNV-GL. Technology Outlook 2030, Safer, Smarter, Greener, DNV GL AS, NO-1322. Høvik, Norway (2020)
2. Grigoropoulos, G.: Shipping and sustainable economic development: The case of Greece. In: Sklias, P., Tzifakis, N. (eds.) Greece's. Horizons. Reflecting on the Country's Assets and Capabilities The Konstantinos Karamanlis Institute for Democracy Series on European and International Affairs. Springer. ISSN 1866-1270, Athens (2012)
3. Hatdjibiros, K.: Ecology and Applied Environmental Science. Taylor & Francis Group, LLC (2019). ISBN-13-978-0-367-86776-8
4. UN Global Compact. Global Goals, Ocean Opportunities. Available at https://www.unglobalcompact.org/ (2019)

Ocean Wildlife and Megafauna Protection

Mohammad Zahirul Islam

1 Introduction

The Earth's oceans cover over 70% of the planet's surface, making them the largest habitat on Earth and home to a vast array of life collectively known as ocean wildlife [6]. From the tiniest plankton that form the base of the marine food web to the majestic blue whales, marine ecosystems harbor an extraordinary diversity of species, each playing a crucial role in maintaining the health and balance of these delicate environments [45]. The significance of ocean wildlife cannot be overstated, as these diverse marine organisms provide essential ecosystem services that support life on Earth [25]. Phytoplankton, for instance, play a vital role in producing oxygen through photosynthesis, contributing approximately 50% of the world's oxygen supply [17]. Additionally, ocean wildlife plays a crucial role in regulating climate patterns by sequestering carbon dioxide, a major greenhouse gas responsible for global warming [61]. By absorbing and storing carbon, marine ecosystems help mitigate the impacts of climate change [12]. Moreover, ocean wildlife serves as the foundation of marine food chains, sustaining various marine species and supporting the livelihoods of millions of people worldwide [47]. Fisheries and aquaculture industries heavily rely on healthy marine ecosystems to provide a steady supply of fish and seafood for human consumption [24]. The fishing industry, in particular, is a significant source of livelihood for coastal communities and contributes to food security and economic growth in many regions [57]. In addition to supporting fisheries, marine wildlife-based tourism has emerged as a substantial economic driver in coastal regions. Activities such as whale watching, diving, and ecotourism draw visitors from around the world, contributing to local economies and creating employment opportunities for communities in coastal areas [34].

M. Z. Islam (✉)
Marinelife Alliance, Cox's Bazar, Bangladesh

S. Th. Rassia (ed.), *The Blue Book*, https://doi.org/10.1007/978-3-031-48831-3_2

2 Diversity of Ocean Wildlife

The Earth's oceans encompass a vast and diverse array of life. From the tiniest plankton to the largest marine megafauna, these diverse species play crucial roles in marine ecosystems and contribute to the overall health and balance of the oceans [11]. The oceans are teeming with life, with estimates suggesting that more than one million species inhabit marine environments [45]. This incredible biodiversity is shaped by a variety of factors, including temperature, ocean currents, and available nutrients. The distribution of marine species varies across different ocean regions, leading to unique and diverse ecosystems [53]. This essay explores the incredible diversity of ocean wildlife, focusing on the various species that inhabit marine environments, including the awe-inspiring marine megafauna that capture our imagination and curiosity.

Marine Bacteria and Microorganisms At the base of the marine food web are microscopic organisms such as phytoplankton, responsible for a significant portion of the Earth's oxygen production through photosynthesis. Zooplankton, including tiny animals such as krill, serve as essential food sources for various marine species [16].

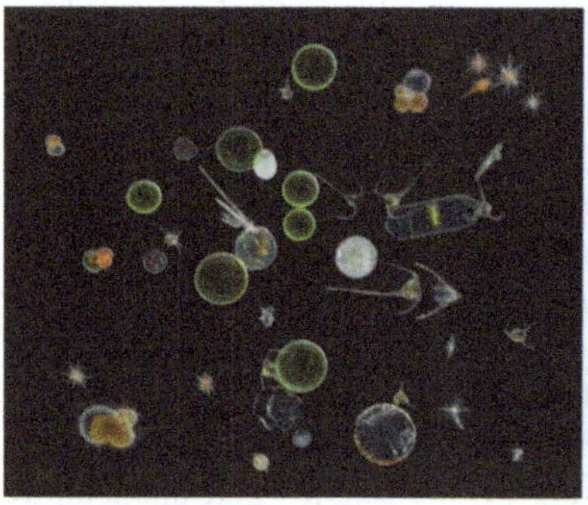

Marine Fish Fish represent a vast and diverse group of marine species, inhabiting virtually all marine environments, from coral reefs to deep-sea trenches. From colorful tropical fish in coral reefs to large pelagic species like tuna and swordfish, marine fish play essential roles in marine ecosystems and are crucial to global food security [47].

Marine Invertebrates Marine environments are home to an astonishing diversity of invertebrates, including mollusks (e.g., clams, squid, octopuses), crustaceans (e.g., crabs, lobsters), and echinoderms (e.g., sea stars, sea urchins). These invertebrates occupy various ecological niches, contributing to ecosystem processes such as nutrient cycling and predation [33].

Marine Reptiles Marine reptiles, such as sea turtles and marine iguanas, have adapted to life in the oceans and are known for their migratory behaviors across vast distances. Sea turtles, in particular, are iconic marine species and play critical roles in maintaining the health of marine ecosystems by grazing on seagrass beds and contributing to nutrient cycling [36].

Diversity of Marine Mammals Marine mammals encompass a wide range of species, each adapted to different marine environments and ecological niches. Some of the major groups of marine mammals include: (a) Cetaceans (Whales, Dolphins, and Porpoises): Cetaceans are perhaps the most iconic marine mammals, and they can be further divided into two subgroups – toothed whales (e.g., killer whales and sperm whales) and baleen whales (e.g., humpback whales and blue whales); (b) Pinnipeds (Seals, Sea Lions, and Walruses): Pinnipeds are marine mammals with flippers, adapted to both land and water. They are found in various habitats, from the icy waters of the Arctic to the warmer regions; (c) Sirenians (Manatees and Dugongs): Sirenians are large, slow-moving marine mammals found in tropical and subtropical waters. They are herbivores and play essential roles in maintaining seagrass ecosystems.

The diversity of ocean wildlife is astounding, with countless species inhabiting marine environments and contributing to the functioning and balance of marine ecosystems. From the smallest microorganisms to the magnificent marine megafauna, each species has a vital role to play in sustaining the health and richness of our oceans. Understanding and preserving this diversity is crucial for the well-being of our planet and future generations to come.

3 Marine Megafauna: The Majesty of the Ocean Giants

Marine megafauna refers to large marine species that capture the imagination of people worldwide due to their impressive size and charismatic nature. These creatures often serve as flagship species for marine conservation efforts, drawing attention to the broader issues of ocean wildlife protection. These ocean giants serve as flagship species for marine conservation efforts, drawing attention to the broader issues of ocean wildlife protection and the urgent need for sustainable marine management [2]. Their immense significance to marine ecosystems cannot be overstated, as they play essential roles in maintaining ecological balance and contributing to the overall health of our oceans. Marine megafauna is captivating and diverse group of large marine species roam the world's oceans. These magnificent creatures capture the imagination of people worldwide, evoking awe and wonder with their impressive size, charismatic behaviors, and pivotal ecological roles. Representing some of the most iconic inhabitants of the seas, marine megafauna includes a wide array of species, ranging from mighty whales and graceful dolphins to formidable sharks and ancient marine turtles. Some notable marine megafauna include:

Whales Whales are among the most iconic marine megafauna, ranging from the enormous blue whale, the largest animal on Earth, to smaller species like humpback and orca whales. These majestic creatures undertake long migrations, sing complex songs, and play vital roles in marine ecosystems as top predators. Whales, the gentle leviathans of the deep, are among the most well-known marine megafauna. From the awe-inspiring blue whale, the largest animal on Earth, to the acrobatic humpback and the intelligent orca, whales possess extraordinary abilities and behaviors that captivate researchers and nature enthusiasts alike [51]. Whales, the largest creatures to roam the oceans, are among the most majestic and enigmatic beings on the Earth. These magnificent marine mammals belong to the order Cetacea and are divided into two main suborders: baleen whales (Mysticeti) and toothed whales (Odontoceti). Whales hold a special place in human culture and history, inspiring awe and fascination for centuries. From their grand size and impressive migrations to their complex social behaviors and haunting songs, whales continue to captivate the hearts and minds of people worldwide. Baleen whales are filter-feeders, using specialized plates called baleen to sift small prey, such as krill and small fish, from the water. Among the most iconic of the baleen whales are the humpback whales (Megaptera novaeangliae) known for their spectacular acrobatic displays and haunting songs that echo across the vast ocean. Toothed whales, on the other hand, are active hunters, relying on echolocation to locate and capture their prey, which can range from fish and squid to even other marine mammals. One of the most celebrated toothed whales is the sperm whale (Physeter macrocephalus), famous for its massive size and association with the classic novel "Moby-Dick" [65].

The conservation of whales has been a focal point for marine wildlife protection and international cooperation. Historically, whales faced significant threats from commercial whaling, leading to severe population declines in many species. However, global efforts, such as the International Whaling Commission's moratorium on commercial whaling, have allowed some whale populations to recover [35]. Despite this progress, whales still face numerous challenges in the modern era. Threats include ship strikes, entanglement in fishing gear, habitat degradation, and the impacts of climate change, such as ocean acidification and shifting food availability [66]. Whales are vital to marine ecosystems, serving as keystone species that help regulate prey populations and distribute nutrients throughout the ocean. As large-bodied predators, they play a crucial role in maintaining the balance of marine food webs.

In conclusion, whales stand as iconic guardians of the deep, embodying the splendor and mysteries of the oceans. Their conservation remains essential not only to protect these awe-inspiring creatures but also to preserve the health and integrity of marine ecosystems as a whole. By working together on international conservation initiatives and promoting sustainable practices, we can ensure that whales continue to grace the oceans for generations to come.

3.1 Ocean Dolphins: Intelligent and Social Marine Mammals

Dolphins are marine mammals, highly intelligent, known for their playful antics, acrobatic behaviors and social dynamics known for their and playful interactions. They are found in oceans worldwide and are often seen swimming alongside boats and interacting with humans. They are communicative, display intricate behaviors, and form complex social structures, making them a fascinating subject of scientific study. The social structures of dolphin communities are intricate and dynamic, forming tight-knit pods that can consist of a few individuals to hundreds of members. Within these pods, dolphins engage in cooperative behaviors, such as hunting and protecting their young, emphasizing their strong sense of social bonds [9]. Ocean dolphins, a fascinating group of marine mammals, captivate our hearts with their intelligence, playful behavior, and close-knit social structures. These charismatic creatures are members of the Delphinidae family, and they are found in oceans worldwide, from the coastal waters to the open sea. Dolphins have long been a subject of scientific fascination and admiration among humans, owing to their remarkable cognitive abilities and complex communication. One of the most well-known and beloved species is the bottlenose dolphin (*Tursiops truncatus*). Bottlenose dolphins are found in various marine habitats, including coastal and offshore waters, and are often spotted swimming alongside boats, displaying their acrobatic skills and inquisitive nature [32]. Dolphins are highly intelligent animals, exhibiting problem-solving skills, self-awareness, and sophisticated communication abilities. Their large brains relative to body size suggest their remarkable cognitive capacity, making them a subject of extensive research in the fields of neuroscience and animal behavior [43]. Communication among dolphins is diverse and complex, involving a combination of vocalizations, body postures, and gestures. Each pod develops its unique repertoire of whistles and clicks, enabling them to identify and communicate with one another [44].

Sharks Sharks, the apex predators of the seas, are crucial in maintaining the delicate balance of marine food webs. Often misunderstood due to sensationalist portrayals in popular media, most shark species are not a threat to humans. Instead, they are key players in regulating marine ecosystems and ensuring ecological stability [14]. The whale shark (*Rhincodon typus*) is a magnificent and awe-inspiring creature, often referred to as the gentle giant of the ocean. It is the largest known extant fish species, captivating the hearts of marine enthusiasts and researchers alike. This magnificent fish can reach lengths of up to 40 feet (12 m) or more, and its unique appearance and peaceful nature have made it a favorite subject for underwater encounters and ecotourism. The whale shark's distinctive appearance is characterized by its massive size, broad flattened head, and mouth that stretches nearly 5 feet wide (1.5 m). Its body is adorned with a striking pattern of white spots and stripes against a dark background, which is unique to each individual, allowing researchers to identify and track them [54]. Despite its imposing size, the whale shark is a filter feeder, primarily consuming plankton, small fish, and other microscopic organisms. Using its wide mouth like a filter, it gently sieves water and traps tiny prey with its specialized gill rakers, ensuring minimal impact on other marine life. Whale sharks are widely distributed in tropical and warm temperate seas around the world, and they are known for their impressive migrations across vast oceanic distances [56]. These migratory patterns have drawn attention to the need for international collaboration in their conservation, as their movements often span multiple countries' waters.

Whale shark (*Rhincodon typus*) roaming in the Maldivian waters (Credit: Sebastian Pena Lambarri)

Conservation efforts for whale sharks focus on understanding their migration patterns, behavior, and population dynamics. Their protected status is crucial to safeguarding their habitat and ensuring sustainable tourism practices that respect the natural behavior of these gentle giants [48].

While the whale shark is listed as "Endangered" on the International Union for Conservation of Nature (IUCN) Red List, the recognition of their ecological significance and the implementation of responsible conservation measures provide hope for their continued existence and the preservation of their role in maintaining the health and balance of marine ecosystems. In conclusion, the whale shark stands as a magnificent ambassador for the conservation of marine biodiversity. Its peaceful demeanor, awe-inspiring size, and unique ecological role make it a symbol of the splendor and importance of marine life. Protecting and understanding these gentle giants not only preserves their species but also contributes to the broader conservation of our oceans and the incredible diversity of life they harbor.

3.2 Sea Turtle: Guardians of the Ocean

Marine Turtles Marine turtles are ancient creatures that have been around for millions of years. Sea turtles are ancient mariners that have gracefully navigated the world's oceans for millions of years. These captivating reptiles hold a special place in the hearts of people worldwide, inspiring awe and admiration with their incredible journeys and vital ecological roles. As guardians of the ocean, sea turtles play a crucial part in maintaining the health and balance of marine ecosystems. Marine turtles, with their ancient lineage dating back millions of years, epitomize resilience and endurance. Undertaking remarkable migrations across vast oceanic distances, these creatures contribute significantly to the health of seagrass beds and coral reefs, vital ecosystems that support a wealth of marine life [30]. Seven species of sea turtles exist, and each possesses unique characteristics and adaptations that allow them to thrive in various marine habitats [63]. These species include the loggerhead, green, leatherback, hawksbill, Kemp's ridley, olive ridley, and flatback turtles. Their impressive migrations across vast oceanic distances have fascinated scientists and conservationists, making them emblematic of endurance and resilience. Sea turtles are vital to marine ecosystems as they perform several essential functions. For instance, they play a crucial role in the health of seagrass beds and coral reefs. As herbivores, some sea turtle species graze on seagrass, helping to maintain the beds and promoting the growth of healthy vegetation [31]. In turn, healthy seagrass beds support a wealth of marine life, providing nursery grounds and shelter for various species. Additionally, sea turtles contribute to nutrient cycling in marine environments. When they come ashore to nest, they dig nests and lay eggs, which act as nutrient-rich fertilizer for the beach and surrounding dunes [49]. This process helps maintain the health of coastal ecosystems and supports the diverse flora and fauna that depend on these habitats.

Seabirds Although not as large as some other marine megafauna, seabirds are charismatic and highly adapted to life at sea. They undertake extensive migrations, often traveling thousands of kilometers to find food. The enchanting seabirds, though not as large as some of their marine counterparts, are charismatic avian residents of the world's oceans. Adapted to life at sea, they undertake incredible journeys, covering thousands of kilometers in search of food, enriching marine ecosystems through their feeding behaviors [64].

Large Seabird Australasian gannet (*Morus serrator*), inhabits Coasts of New Zealand, Victoria, and Tasmania

Protecting marine megafauna is a critical endeavor, as their survival is intertwined with the well-being of marine ecosystems and the sustainability of human activities that rely on healthy oceans. Conservation efforts aimed at preserving these awe-inspiring creatures contribute to the preservation of marine biodiversity, the enhancement of ecosystem resilience, and the protection of the very lifeblood of our planet – the oceans. They play essential roles in marine ecosystems, contributing to biodiversity, nutrient cycling, and ecosystem stability. However, many of these species face various threats, including habitat degradation, pollution, climate change, and overfishing. Conservation efforts focused on protecting marine megafauna not only help safeguard these iconic species but also contribute to the preservation of marine ecosystems and the overall health of our oceans.

4 Importance of Ocean Wildlife

Ocean wildlife is of great ecological, economic, and cultural importance. Vital for marine food webs, marine species, including phytoplankton and top predators, form interconnected relationships, regulating populations and ecosystem health.

This interdependence highlights their crucial roles in preserving balance and population dynamics within marine ecosystems [13]. Coral reefs, often called the "rainforests of the sea," are home to a remarkable diversity of marine life, providing shelter, breeding grounds, and nurseries for countless species that support an astounding diversity of marine life [5]. Moreover, the ocean supports crucial ecosystem services such as carbon sequestration, oxygen production, and climate regulation, playing a vital role in mitigating climate change. Ocean wildlife is critical to the functioning of marine ecosystems and sustains life on Earth in various ways. These vibrant ecosystems provide crucial habitat and shelter for a myriad of species, including fish, crustaceans, and mollusks. Coral reefs also act as natural barriers, protecting coastlines from erosion and storm surges, benefiting both marine and human communities [1, 21]. Economically, ocean wildlife is a significant resource for fisheries, aquaculture, and tourism industries, providing sustenance and livelihoods for millions of people worldwide. Beyond ecological significance, ocean wildlife plays a pivotal role in sustaining human livelihoods [58]. Additionally, marine life holds cultural value, forming the foundation of traditional practices, art, and spiritual beliefs in many societies [50]. For instance, phytoplankton, tiny photosynthetic organisms, are responsible for approximately half of the world's oxygen production, making them vital to global atmospheric processes [18]. Coastal communities heavily rely on marine resources for food security and economic prosperity. Fisheries and aquaculture industries heavily rely on marine resources for food and economic prosperity [8]. Coastal communities worldwide depend on fishing as a primary source of income and sustenance. Moreover, marine wildlife-based tourism, such as whale watching and diving, generates significant economic revenue in many regions [46]. In conclusion, ocean wildlife represents an invaluable natural heritage that sustains marine ecosystems, supports livelihoods, and provides essential ecosystem services to humanity. The threats posed by human activities demand urgent and concerted conservation efforts to protect and preserve this fragile biodiversity. By implementing effective measures such as marine protected areas, sustainable fisheries management, pollution reduction, and climate change mitigation, we can ensure a sustainable future for ocean wildlife and the ecosystems they inhabit. It is our collective responsibility to safeguard these majestic and diverse inhabitants of the sea for generations to come.

5 Threats to Ocean Wildlife

Ocean wildlife faces a multitude of threats that endanger marine species and disrupt the delicate balance of marine ecosystems. These threats stem from human activities, climate change, and natural factors, collectively posing significant challenges to the preservation of marine biodiversity and ecological stability. The rich diversity of ocean wildlife, including marine megafauna, is facing unprecedented threats from human activities, including overfishing, habitat destruction, pollution, and climate change. These threats not only impact individual species but also disrupt entire marine ecosystems, with far-reaching consequences for human livelihoods and well-being.

1. *Overfishing*, the unsustainable harvesting of fish stocks, poses a grave threat to marine ecosystems and global food security. The practice of catching fish at a rate faster than they can reproduce leads to depleted populations, disrupts marine food webs, and compromises ecological balance [38]. Over 33% of global fish stocks are overfished, risking irreversible damage to marine biodiversity and coastal communities that heavily depend on fisheries for livelihoods and sustenance [19]. Implementing sustainable fishing practices, such as catch limits and ecosystem-based management, is essential to mitigate the impacts of overfishing and safeguard marine resources for future generations.
2. *Ocean habitat loss* poses a significant threat to marine biodiversity, with human activities, such as coastal development, destructive fishing practices, and pollution, leading to the degradation and destruction of vital marine ecosystems. Coral reefs, seagrass beds, and mangroves are among the most affected habitats. This loss of habitat negatively impacts numerous marine species, disrupts ecological balance, and reduces resilience to climate change impacts. Urgent conservation efforts are required to mitigate the consequences of ocean habitat loss and protect marine ecosystems.
3. *Ocean pollution* poses a severe threat to marine ecosystems, wildlife, and human health. Human activities, such as industrial discharges, agricultural runoff, and plastic waste, introduce pollutants into the marine environment [37]. This pollution results in habitat degradation, harming marine species through ingestion, entanglement, and disruption of natural behaviors. Moreover, toxic substances accumulate in the food chain, potentially impacting human health through seafood consumption. Urgent action is needed to address ocean pollution and protect the delicate balance of marine ecosystems [21].
4. *Climate change* poses significant threats to ocean wildlife, including rising sea temperatures, ocean acidification, and altered marine habitats. Warmer waters disrupt marine species' life cycles, while ocean acidification harms organisms with calcium carbonate shells or skeletons. Additionally, changing environmental conditions may force marine species to migrate, impacting their natural ecosystems. These threats threaten the survival and ecological balance of marine biodiversity [20, 28].
5. *Bycatch*: Threats to ocean wildlife due to bycatch result from the unintentional capture of nontarget species in fishing gear, leading to mortality and ecological imbalances. Bycatch affects marine biodiversity and jeopardizes the conservation status of vulnerable species such as dolphins, sea turtles, and seabirds [23]. The incidental catch can have severe implications for marine ecosystems and disrupt food webs. Effective mitigation measures and sustainable fishing practices are essential to reduce the impact of bycatch on ocean wildlife and maintain healthy marine ecosystems.
6. *Invasive species* pose a significant threat to ocean wildlife by disrupting natural ecosystems and outcompeting native species. These non-native organisms, intro-

Whale shark (*Rhincodon typus*) incidentally catch in Bangladesh offshore fishing activities. The shark at right side luckily rescued, helped to gain breath and released successfully by Marinelife Alliance's marine megafauna conservation program, 2014 in Bangladesh. (Credit: Marinelife Alliance)

duced through ballast water discharge, aquaculture, or accidental releases, can rapidly spread and cause ecological imbalances [55]. Invasive species such as lionfish, zebra mussels, and green crabs can devastate local marine populations, alter food webs, and damage critical habitats. Their proliferation can lead to declines in biodiversity and diminish the resilience of marine ecosystems against other threats. Preventive measures and early detection are crucial to mitigating the impacts of invasive species on ocean wildlife.

7. *Noise pollution* in the ocean poses a significant threat to marine wildlife. Anthropogenic noise from ship traffic, seismic exploration, and military activities disrupts marine animals' communication, navigation, and feeding behaviors, leading to stress and reduced reproductive success. Whales, dolphins, and other marine species that rely on sound for crucial life functions are particularly vulnerable. Noise pollution can also mask important environmental cues and affect the distribution of marine species. To mitigate these impacts, stricter regulations on underwater noise emissions and the establishment of marine protected areas are crucial [15, 31].
8. *Ocean acidification* poses a significant threat to ocean wildlife, especially marine organisms with calcium carbonate shells or skeletons, including corals, mollusks, and some planktonic species. As the ocean absorbs excess carbon dioxide from the atmosphere, its pH decreases, hindering the ability of these organisms to build and maintain their structures. This disrupts marine food chains, alters ecosystems, and diminishes biodiversity [20]. The impact of acidification on marine life highlights the urgent need for global efforts to mitigate carbon emissions and safeguard the delicate balance of marine ecosystems.

6 Conservation of Ocean Wildlife and Marine Megafauna

The conservation of ocean wildlife and marine megafauna is of utmost importance to maintain the health of marine ecosystems and preserve the rich biodiversity that inhabits the world's oceans. Marine megafauna refers to large marine animals, such as whales, dolphins, sharks, sea turtles, and seals, which play crucial ecological roles and capture the public's imagination due to their size, beauty, and charisma. However, these magnificent creatures face a myriad of threats, primarily driven by human activities. Effective conservation measures are essential to ensure the survival of marine megafauna and the integrity of marine ecosystems for future generations. Conservation of ocean wildlife, especially marine megafauna, is essential for the health and resilience of marine ecosystems. By addressing the threats faced by marine megafauna and implementing effective conservation strategies, we can ensure the survival of these majestic creatures and safeguard the biodiversity and ecological integrity of our oceans. Collaboration between governments, conservation organizations, scientists, and the public is crucial to achieving these conservation goals and creating a sustainable future for marine wildlife. To safeguard marine megafauna and promote ocean wildlife conservation, several strategies have been implemented:

1. *Marine Protected Areas (MPA)*

MPAs are crucial tools in marine conservation, providing essential refuges for marine megafauna and safeguarding critical habitats. By limiting human activities within designated areas, MPAs reduce the impact of fishing, coastal development, and pollution on marine wildlife [26]. These protected zones serve as safe havens for species like whales, dolphins, sharks, and sea turtles, allowing them to thrive and maintain ecological balance. Establishing MPAs is vital to preserving marine biodiversity and ensuring the long-term sustainability of marine megafauna populations. Furthermore, establishing and effectively managing MPA is one of the most effective strategies to protect marine biodiversity. MPAs offer refuge for marine species, allowing populations to recover and thrive, while also protecting critical habitats and breeding grounds [27, 41, 42]. To safeguard the invaluable natural heritage of ocean wildlife and ensure the health and resilience of marine ecosystems, urgent conservation efforts are required. By designating and properly managing MPAs, we can contribute significantly to the preservation of marine megafauna and the overall well-being of our oceans.

2. *Bycatch Mitigation*

is a crucial aspect of ocean wildlife protection, especially concerning marine megafauna conservation. Bycatch refers to the unintentional capture of nontarget species in fishing gear, and it poses a significant threat to marine megafauna populations worldwide. However, implementing technological and policy measures can effectively reduce bycatch and minimize its impact on these vulnerable species. One approach to bycatch mitigation is using more selective fishing gear. By adopting gear modifications or alternative fishing methods, fishermen can target specific species while reducing the capture of nontarget species like marine megafauna. For instance, using circle hooks in longline fisheries has been shown to reduce the incidental catch of sea turtles and marine mammals [39, 40]. Additionally, the use of acoustic deterrent devices can help reduce interactions between marine megafauna and fishing gear, thereby mitigating bycatch incidents. Policy measures are equally critical for effective bycatch mitigation. Governments and fisheries management authorities can establish and enforce regulations that require the use of selective fishing gear, set limits on bycatch, and impose penalties for noncompliance. Seasonal closures of certain areas during peak migration or breeding seasons can also reduce interactions between fishing activities and marine megafauna populations. By addressing bycatch through a combination of technological advancements and well-designed policies, we can significantly contribute to ocean wildlife protection, particularly the conservation of marine megafauna. Such mitigation efforts are essential for maintaining marine biodiversity, ensuring sustainable fisheries, and safeguarding the delicate balance of marine ecosystems.

3. *Sustainable Fisheries Management*

plays a crucial role in the conservation of marine megafauna. By adopting sustainable fishing practices, fisheries can mitigate the impact of fishing activities on marine wildlife, including large and vulnerable species such as marine megafauna. Implementing catch limits is an essential component of sustainable fisheries management. Setting scientifically based catch limits ensures that fish populations are not overexploited, allowing them to recover and maintain healthy population levels [10]. Implementing fishing quotas, employing selective fishing gear, and promoting responsible fishing practices can help reduce overfishing and mitigate the impacts of bycatch. By preventing overfishing, catch limits help safeguard the food supply and ecological interactions of marine megafauna, which depend on balanced marine food webs. Promoting ecosystem-based management is another key aspect of sustainable fisheries management. Rather than focusing solely on individual species, ecosystem-based management considers the entire marine ecosystem and its interactions. This approach recognizes the interconnectedness of marine species and habitats, acknowledging that the health of marine megafauna is closely tied to the overall ecosystem. By protecting and managing the entire ecosystem, the conservation of marine megafauna is better integrated into broader conservation efforts. These sustainable fishing practices not only benefit marine megafauna but also support the long-term viability of fishing industries and coastal communities that rely on healthy marine ecosystems for their livelihoods [58]. By ensuring the responsible and balanced use of marine resources, sustainable fisheries management contributes to the preservation of marine biodiversity and the well-being of both marine wildlife and human communities.

4. *Habitat Restoration*

Restoration plays a crucial role in marine conservation, particularly for marine megafauna such as whales, dolphins, and sea turtles. Restoring and protecting vital marine habitats, such as coral reefs and seagrass beds, creates critical environments where these majestic creatures can thrive. Coral reefs are among the most diverse and productive marine ecosystems, providing essential shelter, feeding grounds, and breeding areas for numerous marine species [3]. However, coral reefs face severe threats, including coral bleaching due to rising sea temperatures and ocean acidification. By implementing conservation efforts, such as restoring damaged coral reefs and reducing pollution, we can help create a more favorable environment for marine megafauna that rely on these ecosystems for survival. Seagrass beds are equally important habitats for marine megafauna. These underwater meadows provide vital feeding grounds for sea turtles, dugongs, and manatees, while also acting as nursery areas for juvenile fish [62]. Restoring and protecting seagrass habitats can enhance the resilience of marine megafauna populations, providing them with the necessary resources for successful reproduction and survival. Habitat restoration initiatives should involve collaborative efforts among governments, conservation organizations, scientists, and local communities. Strategies may include reducing coastal development and pollution, implementing fishing regulations, and creating marine protected areas to ensure the long-term viability of marine megafauna and the preservation of marine biodiversity.

5. *Plastic Pollution*

Ocean pollution poses a significant threat to marine megafauna, including whales, dolphins, sea turtles, and seals, as these large marine animals often mistake plastic debris for food or become entangled in plastic waste. Implementing measures to reduce plastic waste and improve waste management is essential to mitigate the impact of plastic pollution on these iconic creatures. One effective approach is to implement policies and regulations to reduce the production and use of single-use plastics, such as plastic bags and straws. Many countries and cities have already taken steps to ban or restrict the use of single-use plastics to curb their proliferation in the environment [22]. Promoting recycling and waste reduction programs is another crucial strategy. Encouraging proper waste disposal and recycling practices can help prevent plastic waste from entering waterways and eventually reaching the oceans. Public awareness campaigns about the importance of reducing plastic consumption and the consequences of improper waste disposal can also contribute to behavior change [37]. Moreover, encouraging the development of sustainable alternatives to plastic can help reduce the demand for traditional plastics. Biodegradable or compostable materials can be viable alternatives for certain applications, reducing the persistence of plastic waste in the environment [60]. Addressing ocean pollution requires a multifaceted approach, including reducing single-use plastics, implementing waste management strategies, and promoting recycling.

6. *Climate Change Mitigation*:

Climate change mitigation and adaptation measures are vital to protect ocean wildlife from the impacts of warming temperatures and ocean acidification [7]. Transitioning to renewable energy sources and reducing greenhouse gas emissions are essential to limit further climate change effects on marine ecosystems [52]. Addressing climate change through global efforts to reduce greenhouse gas emissions is crucial to protect marine megafauna and their habitats. Climate change is one of the most significant threats facing marine ecosystems and wildlife, including marine megafauna. The burning of fossil fuels, deforestation, and industrial activities release greenhouse gases, such as carbon dioxide (CO_2) and methane, into the atmosphere, leading to the greenhouse effect. This effect traps heat within the Earth's atmosphere, resulting in rising global temperatures and altering oceanic conditions. The warming of the oceans affects marine megafauna in various ways. It can lead to shifts in the distribution and availability of prey species, disrupt breeding and migration patterns, and cause changes in the timing of critical life events, such as mating and calving [28]. Additionally, ocean acidification, a consequence of increased CO_2 absorption by the oceans, negatively impacts marine species with calcium carbonate shells, including corals and some marine mollusks [20]. By reducing greenhouse gas emissions on a global scale, we can mitigate the impacts of climate change on marine megafauna and their habitats. Transitioning to renewable energy sources, promoting energy efficiency, and adopting sustainable land-use practices are essential steps in mitigating climate change. The Paris Agreement, a landmark international climate accord, aims to unite nations in combatting climate change and limiting global temperature rise. The implementation of ambitious emission reduction targets and policies is critical to achieving the agreement's goals and safeguarding marine ecosystems and wildlife.

- *Research and Education*:

Research and education play a pivotal role in the conservation of ocean wildlife, particularly marine megafauna. Through scientific research, we gain valuable insights into the intricate ecology, behavior, and migratory patterns of marine megafauna, enabling informed conservation strategies and the identification of crucial habitats and migration corridors [4]. Understanding the life cycles and behaviors of these magnificent creatures helps guide efforts to protect their populations and preserve their habitats. In combination, research and education create a powerful synergy that drives positive change for ocean wildlife. Robust scientific research informs evidence-based conservation strategies, while education empowers individuals to become advocates for marine megafauna protection, fostering a collective commitment to preserving the awe-inspiring creatures that inhabit our oceans. Public education and outreach are equally crucial components of effective conservation. Raising awareness about the significance of marine megafauna conservation fosters a sense of responsibility among communities, individuals, and policymakers. Educational initiatives inform the public about the threats facing marine megafauna and highlight the broader ecological implications of their preservation. By engaging people of all ages and backgrounds, education promotes sustainable behaviors and encourages support for conservation policies and actions [38]. Public awareness and education are powerful tools in promoting ocean wildlife conservation. Engaging communities and stakeholders in conservation initiatives can foster a sense of stewardship toward marine environments and encourage responsible behavior toward ocean resources.

References

1. Agusti, S., Barbier, E., Britten, G.L., Castilla, J.C., Duarte, C.M., Gattuso, J.P., et al.: Rebuilding marine life. Nature. **580**(7801), 39–51 (2020)

2. Barnett, A., Braccini, M., Payne, N.: Meta-analysis of shark attack: a global perspective. Mar. Policy. **37**, 170–173 (2012)
3. Bellwood, D.R., Pratchett, M.S., Morrison, T.H., Gurney, G.G., Hughes, T.P., Álvarez-Romero, J.G., et al.: Coral reef conservation in the Anthropocene: confronting spatial mismatches and prioritizing functions. Biol. Conserv. **236**, 604–615 (2019)
4. Block, B.A., Jonsen, I.D., Jorgensen, S.J., Winship, A.J., Shaffer, S.A., Bograd, S.J., et al.: Tracking apex marine predator movements in a dynamic ocean. Nature. **475**(7354), 86–90 (2011)
5. Burke, L., Reytar, K., Spalding, M., Perry, A.: Reefs at Risk Revisited. World Resources Institute, Washington, DC (2019)
6. Carpenter, K.E., et al.: One-third of reef-building corals face elevated extinction risk from climate change and local impacts. Science. **321**(5888), 560–563 (1998)
7. Cheung, W.W., Lam, V.W., Sarmiento, J.L., Kearney, K., Watson, R., Zeller, D., & Pauly, D.: Large-scale redistribution of maximum fisheries catch potential in the global ocean under climate change. Global Change Biol. **16**(1), 24–35 (2013)
8. Cinner, J.E., Aswani, S., Basurto, X.: Human dimensions of coral reef social–ecological systems. Ecol. Soc. **23**(1), 14 (2018)
9. Connor, R.C., Mann, J., Tyack, P.L., Whitehead, H.: Social evolution in toothed whales. Trends Ecol. Evol. **15**(4), 148–153 (2000)
10. Costello, C., Ovando, D., Hilborn, R., Gaines, S.D., Deschenes, O., Lester, S.E.: Status and solutions for the world's unassessed fisheries. Sci. Adv. **2**(8), e1500933 (2016)
11. Dayton, P.K., Thrush, S.F., Agardy, M.T., Hofman, R.J.: Environmental effects of marine fishing. Aquat. Conserv. Mar. Freshwat. Ecosyst. **8**(3), 337–354 (1998)
12. Duarte, C.M., Middelburg, J.J., Caraco, N.: Major role of marine vegetation on the oceanic carbon cycle. Biogeosciences. **1**, 173–180, (2005)
13. Duffy, J.E., Lefcheck, J.S., Stuart-Smith, R.D., Navarrete, S.A., Edgar, G.J.: Biodiversity enhances reef fish biomass and resistance to climate change. Proc. Natl. Acad. Sci. **116**(22), 10786–10791 (2019)
14. Dulvy, N.K., Fowler, S.L., Musick, J.A., Cavanagh, R.D., Kyne, P.M., Harrison, L.R., et al.: Extinction risk and conservation of the world's sharks and rays. elife. **3**, e00590 (2014)
15. Erbe, C., Reichmuth, C., Cunningham, K., Lucke, K., Dooling, R.: Communication masking in marine mammals: a review and research strategy. Mar. Pollut. Bull. **103**(1–2), 15–38 (2016)
16. Falkowski, P.G., Barber, R.T., Smetacek, V.: Biogeochemical controls and feedbacks on ocean primary production. Science. **281**(5374), 200–206 (1998)
17. Falkowski, P.G., et al.: The global carbon cycle: a test of our knowledge of earth as a system. Science. **290**(5490), 291–296 (2003)
18. Falkowski, P.G., Fenchel, T., Delong, E.F.: The microbial engines that drive Earth's biogeochemical cycles. Science. **320**(5879), 1034–1039 (2020)
19. FAO: The State of World Fisheries and Aquaculture 2020. Food and Agriculture Organization of the United Nations, Rome (2021)
20. Gattuso, J.P., Magnan, A., Bopp, L., Cheung, W.W.L., Duarte, C.M., Hinkel, J., et al.: Contrasting futures for ocean and society from different anthropogenic CO2 emissions scenarios. Science. **349**(6243), aac4722 (2015)
21. Gattuso, J.P., Magnan, A., Bopp, L., Cheung, W.W.L., Duarte, C.M., Hinkel, J., et al.: Ocean solutions to address climate change and its effects on marine ecosystems. Front. Mar. Sci. **6**, 537 (2020)
22. Geyer, R., Jambeck, J.R., Law, K.L.: Production, use, and fate of all plastics ever made. Sci. Adv. **3**(7), e1700782 (2017)
23. Gilman, E., Brothers, N., Kobayashi, D., Shwiff, S.: Bycatch in small-scale fisheries: a global review. Front. Mar. Sci. **7**, 191 (2020)
24. Golden, C.D., et al.: Nutrition: fall in fish catch threatens human health. Nature. **534**(7607), 317–320 (2016)
25. Halpern, B.S., et al.: A global map of human impact on marine ecosystems. Science. **319**(5865), 948–952 (2008)

26. Halpern, B.S., Frazier, M., Afflerbach, J., Lowndes, J.S., Micheli, F., O'Hara, C., et al.: Recent pace of change in human impact on the world's ocean. Sci. Rep. **10**(1), 1–8 (2020)
27. Harrison, H.B., et al.: The multi-functional role of reef-building corals: a review. Coral Reefs. **28**(2), 385–406 (2019)
28. Hoegh-Guldberg, O., Beal, D., Chaudhry, T., Elvidge, E., Fitzsimmons, C., Frank, G., et al.: Reviving the Ocean Economy: The Case for Action—2017. WWF International, Gland (2017)
29. Hatch, L.T., Clark, C.W., Van Parijs, S.M., Frankel, A.S.: Marine noise pollution – increasing recognition but need for more practical action. J. Appl. Ecol. **54**(5), 1397–1402 (2017)
30. Hays, G.C., Ferreira, L.C., Sequeira, A.M., Meekan, M.G., Duarte, C.M., Bailey, H., et al.: Key questions in marine megafauna movement ecology. Trends Ecol. Evol. **32**(11), 835–846 (2017)
31. Heithaus, M.R., Wirsing, A.J., Thomson, J.A., Burkholder, D.A.: A review of lethal and non-lethal effects of predators on adult marine turtles. J. Exp. Mar. Biol. Ecol. **356**(1–2), 43–51 (2008)
32. Herzing, D.L.: The life history of free-ranging Atlantic spotted dolphins (Stenella frontalis) in The Bahamas. Mar. Mamm. Sci. **29**(3), E221–E235 (2013)
33. Hickman, C.P.: Integrated Principles of Zoology. McGraw-Hill, Boston (1998)
34. Hoyt, E.: Whale Watching 2001: Worldwide Tourism Numbers, Expenditures, and Expanding Socioeconomic Benefits, 158 p. International Fund for Animal Welfare, Yarmouth Port (2001)
35. IWC (International Whaling Commission). Schedule of past whaling catch limits and current quotas. Retrieved from https://iwc.int/catches (2020)
36. Jackson, J.B., Kirby, M.X., Berger, W.H., Bjorndal, K.A., Botsford, L.W., Bourque, B.J., et al.: Historical overfishing and the recent collapse of coastal ecosystems. Science. **293**(5530), 629–637 (2001)
37. Jambeck, J.R., Geyer, R., Wilcox, C., Siegler, T.R., Perryman, M., Andrady, A., et al.: Plastic waste inputs from land into the ocean. Science. **347**(6223), 768–771 (2015)
38. Lemahieu, A., Cisneros-Montemayor, A.M., Ota, Y., Sumaila, U.R.: Global wildlife tourism and carbon emissions: potential geospatial conflicts and co-benefits. J. Travel Res. **57**(6), 709–722 (2018)
39. Lewison, R.L., Freeman, S.A., Crowder, L.B., Freeman, S.A.: Quantifying the effects of fisheries on threatened species: The impact of pelagic longlines on loggerhead and leatherback sea turtles. Ecol. Lett. **7**, 221–231 (2004). https://doi.org/10.1111/j.1461-0248.2004.00573.x
40. Lewison, R.L, Crowder, L.B, Wallace, B.P., Moore, J.E., Cox, T., Zydelis, R., McDonald, S., DiMatteo, A., Dunn, D.C., Kot, C.Y., Bjorkland, R., Kelez, S., Soykan, C., Stewart, K.R., Sims, M., Boustany, A., Read, A.J., Halpin, P., Nichols, W.J., Safina, C.: Global patterns of marine mammal, seabird, and sea turtle bycatch reveal taxa-specific and cumulative megafauna hotspots. Proc. Nat. Acad. Sci. USA. **111**(14), 5271–5276 (2014). https://doi.org/10.1073/pnas.1318960111
41. Lubchenco, J., Gaines, S.D., Warner, R.R., Airamé, S., Simler, B.: The science of marine reserves. Partner. Interdiscipl. Stud. Coast. Oceans. (2002). Available online: http://www.piscoweb.org/outreach/pubs/reserves/index.html
42. Lubchenco, J. Palumbi, S.R., Gaines, S.D., Andelman, S.: Plugging a hole in the ocean: the emerging science of marine reserves. Ecological Applications **1313**(1), Supplement S3–S7 (2013)
43. Mann, J., Sargeant, B.L., Watson-Capps, J.J., Gibson, Q.A., Heithaus, M.R., Connor, R.C., Patterson, E.M.: Why do dolphins carry sponges? PLoS One. **5**(12), e15016 (2010)
44. McCowan, B., Marino, L., Vance, E., Walke, L., Reiss, D.: Bubble ring play of bottlenose dolphins (*Tursiops truncatus*): implications for cognition. J. Comp. Psychol. **114**(1), 98 (2000)
45. Mora, C., Tittensor, D.P., Adl, S., Simpson, A.G.B., Worm, B.: How many species are there on earth and in the ocean? PLoS Biol. **9**(8), e1001127 (2011)
46. Orams, M.B.: Whale-Watching: Sustainable Tourism and Ecological Management. Cambridge University Press, New York (2019)
47. Pauly, D., Christensen, V., Dalsgaard, J., Froese, R., Torres, F.: Fishing down marine food webs. Science. **279**(5352), 860–863 (1998)

48. Pierce, S.J., Méndez-Jiménez, A., Collins, K., Rosero-Caicedo, M., Monadjem, A., Robinson, D.P., et al.: Global diversity, population dynamics, and hotspots of decline in the world's largest terrestrial vertebrates. PLoS One. **15**(10), e0239690 (2020)
49. Pike, D.A.: Nest site selection by sea turtles: integrating animal behaviour with conservation biology. Endanger. Species Res. **13**(1), 1–10 (2011)
50. Posey, D.A., Balée, W., Coppedge, B.R., Erickson, C.L., Hurley, P.T., Moran, E.F.: Ethnobiology, political ecology, and conservation. Annu. Rev. Anthropol. **28**(1), 315–338 (2019)
51. Rendell, L., Whitehead, H.: Vocal clans in toothed whales. Nature. **423**(6942), 847–847 (2003)
52. Riebesell, U., Bach, L.T., Bellerby, R.G., Bermudez, R.: The role of CO2 in making marine life vulnerable to extinction. Environ. Res. Lett. **10**(7), 1–8 (2015)
53. Roberts, C.M., McClean, C.J., Veron, J.E.N., Hawkins, J.P., Allen, G.R., McAllister, D.E., et al.: Marine biodiversity hotspots and conservation priorities for tropical reefs. Science. **295**(5558), 1280–1284 (2002)
54. Rowat, D., Brooks, K.: A review of the biology, fisheries, and conservation of the whale shark. J. Fish Biol. **80**(5), 1019–1056 (2012)
55. Ruiz, G.M., Fofonoff, P.W., Carlton, J.T., Wonham, M.J., Hines, A.H.: Invasion of coastal marine communities in North America: apparent patterns, processes, and biases. Annu. Rev. Ecol. Evol. Syst. **42**, 427–448 (2011)
56. Sequeira, A.M., Mellin, C., Fordham, D.A., Meekan, M.G., Bradshaw, C.J., McMahon, C.R.: Inferred global connectivity of whale shark Rhincodon typus populations. J. Fish Biol. **92**(2), 831–847 (2018)
57. Sumaila, U.R., et al.: Winners and losers in a world where the high seas is closed to fishing. Sci. Rep. **2**, 952 (2012)
58. Sumaila, U.R., Cheung, W.W.L., Lam, V.W.Y., Pauly, D., Herrick, S.F.: Climate change impacts on the biophysics and economics of world fisheries. Nat. Clim. Chang. **11**(1), 16–20 (2020)
59. Sutherland, W.J., Aveling, R., Brooks, T.M., Clout, M., Dicks, L.V., Fellman, L., et al.: A horizon scan of global conservation issues for 2010. Trends Ecol. Evol. **25**(1), 1–7 (2010b)
60. Tecsa, V., Krystynowicz, A., Guzik, U., Labuzek, S., Jesionowski, T., Bródka, M.: Biodegradable polymers and their blends: perspective for oil pollution remediation on marine environments. Sci. Total Environ. **712**, 136299 (2020)
61. Trisos, C.H., et al.: The projected timing of climate departure from recent variability. Nature. **502**(7470), 183–187 (2020)
62. Unsworth, R.K., McKenzie, L.J., Collier, C.J., Cullen-Unsworth, L.C., Duarte, C.M., Eklöf, J.S., et al.: Global challenges for seagrass conservation. Ambio. **48**(7), 801–815 (2019)
63. Wallace, B.P., DiMatteo, A.D., Hurley, B.J., Finkbeiner, E.M., Bolten, A.B., Chaloupka, M.Y., et al.: Regional management units for marine turtles: a novel framework for prioritizing conservation and research across multiple scales. PLoS One. **5**(12), e15465 (2010)
64. Weimerskirch, H.: Are seabirds foraging for unpredictable resources? Deep-Sea Res. II Top. Stud. Oceanogr. **54**(3–4), 211–223 (2007)
65. Whitehead, H., Weilgart, L., Payne, R.: Echolocation clicks of two free-ranging, oceanic delphinids with different food preferences: false killer whales Pseudorca crassidens and Risso's dolphins Grampus griseus. J. Exp. Biol. **200**(1), 69–83 (1997)
66. Williams, R., Ashe, E., O'Hara, P.D., Robichaud, D.: Sperm whales reduce foraging effort during exposure to 1–2 kH z sonar and killer whale sounds. J. Exp. Biol. **219**(24), 3788–3797 (2016)

Big Data and Sustainable Applications for Healthy and Safe Coastal Cities

Amalia Ntemou, Akrivi Kiousi, and Nasos Grigoropoulos

1 Introduction

A huge amount of data is generated by humans on a daily basis but as businesses transform as more digital data is larger, resulting to more complex data sets, especially from new data sources. Because of the great potential it inherently offers, the above are encapsulated to the concept of big data that has become a very interesting topic, transforming the management, exploration, analysis, and leveraging of data in various domains [1, 2]. Various public and private sector industries generate, store, and analyze big data with an aim to improve the services they provide. This evolution of big data analytics has been supported by the emergence of Internet of Things (IoT). Robust big data management infrastructures provide the ground for managing and processing enormous amount of data, realizing different applications of Big Data analytics in various fields such as healthcare, maritime, banking, retail, computing, space communications etc. [3] Therefore industries, including the healthcare industry, are taking vigorous steps to convert this potential into better services and financial advantages.

2 Big Data in Healthcare

From an economic perspective, healthcare is one of the most significant industries. Over the last decade, total healthcare expenditure has increased from 10% (2010) to 11.3% (2019) of GDP. The recent healthcare crisis (COVID-19) caused an initial

A. Ntemou (✉) · A. Kiousi · N. Grigoropoulos
Netcompany-Intrasoft S.A., Luxembourg, Luxembourg
e-mail: amalia.ntemou@netcompany.com; akrivi.kiousi@netcompany.com; athanasios.grigoropoulos@netcompany.com

S. Th. Rassia (ed.), *The Blue Book*, https://doi.org/10.1007/978-3-031-48831-3_3

increase to a level of 12.1% of GDP by 2021 and today (2023), a decrease of 0.4% has been reported. Over the entire period until 2050, healthcare expenditure increases from 11.3% to 15% of GDP [4].

On the other hand, healthcare is a big-data repository backup by a complex system with varied stakeholders: patients, doctors, hospitals, pharmaceutical companies, and healthcare decision-makers. A vast amount of heterogeneous data, such as demographic, socioeconomic, genomics, pharmaceutical, medical imaging, socio-behavioral, sensor data – vital signs or data from mobile apps, to name a few, are constantly generated. Being aware of a high degree of fragmentation in the health sector, a reasonable question arises – do we know enough about the information we can learn from them? Can we do more and make the most of technology?

The footprint of big Data technologies is already evident in healthcare but more needs to be done. An efficient management, analysis, and interpretation of big data can change the game by opening new avenues for modern healthcare. Infection prevention, medical diagnosis, home care, quantifying lifestyle data in the fitness industry are some indicative examples of the applications of big data in healthcare but mostly big data would allow the implementation of personalized and precise medicine based on personalized information, delivered in real time and tailored to individual patients, which is envisaged in the agenda of many governments and experts.

The opportunities opened up by big data are tremendous and affect various dimensions of healthcare, spanning from the reason of what happened, diagnostic to respond to why it happened, forecasting of future events and able to determine the way it might happen. Huge productivity bottlenecks may be unlocked due to big data technologies, and healthcare systems can be of quality and more accessible [5].

3 Big Data-Driven Crisis Management

Unexpected healthcare situations, such as that of COVID-19 pandemic, create the need to tackle the said challenges effectively and extremely quickly. Unpreparedness for crisis management puts populations and governments in a state of worry due to the uncertainty about the future, virus transmission and health effects, as well as uncertainty about the policies and strategies to mitigate it.

The benefits of big data analytics and its evolution can also be seen in the fact that healthcare sector is driven by data (perfect example is the COVID-19 pandemic), which highlights again the advantage of big data for efficient health care delivery [6].

Recently, different countries have turned their attention to big data and advanced analytics technologies incl. Artificial Intelligence (AI) for crisis management, as humanity and the global economy require innovative solutions to address and monitor the evolutions of communicable-infectious diseases [7, 8].

4 Big Data in Maritime

Digitalization leads to smarter and cost-effective decisions while transforming societies and cities to smart. However, to successfully address challenges, emphasis needs to be placed on many domains and activities and their common aspects.

To be more precise, talking about smart cities does not necessarily mean that it only refers to purely urban areas. Coastal cities, for example, may be impacted by marine activities more than the urban ones per se. In other words, costal and ocean proximate cities can equally be smart, incorporating similar technologies as those implemented in cities.

From IoT-enabled maritime intelligent transportation systems, advanced ship operations driven by AI, to smart systems for infection control, big data has a lot to offer especially from controlling sensors on a ship to perform predictive analysis to avoid delays and improve efficiency and enhance prediction mechanisms. Indicatively, some of the fields that can improve ship operations and management are big data analytics, remote monitoring, carbon-based metrics, preventive and corrective equipment maintenance, forecasting [9] and optimizing operations to chime with ship calls, renewing port assets and ensuring optimum cyber-security (Trelleborg) [10].

5 Big Data Management Infrastructures

It goes without saying that crisis on ships management on ships could also be successful because of Big Data Analytics. In line with EU strategic goals for safe, resilient transport and smart mobility services, Horizon Europe projects, namely, Healthy Ship 4U (HS4U) [11] leverage on the opportunities of Big Data to effectively control infection on large passenger ships. The idea is to build a collaborative digital framework (CDF) for live interaction of humans (crew/passengers) and IoT (sensors, measures, actuators) edge devices on the ship based on co-robotics concepts and Artificial Intelligence mechanisms and models. New smart technologies and products as well as stream handling of big data on INTRA platform providing interoperability of functional, design and training process using ML and DL algorithms will be implemented. And this can only be considered as a starting point for the transformation of ships and transportation systems in general to digital and "healthy."

6 Conclusion

It is evident that industries have significant benefits to earn from implementing big data analytics into their business operations. The potential of big data analysis with respect to the engagement of marine and health sectors is huge and applications are constantly being created. Undoubtedly, big data management infrastructures can realize the vision of sustainable, healthy, and safe coastal cities. However, as technology evolves, nothing remains the same and we have a lot to learn from (big) data.

References

1. Dash, S., Shakyawar, S.K., Sharma, M., Kaushik, S.: Big data in healthcare: management, analysis and future prospects. J. Big Data. **6**(1), 1–25 (2019)
2. Zhang, Y., Hong, J., Chen, S.: Medical big data and artificial intelligence for healthcare. Appl. Sci. **13**(6), 3745 (2023)
3. Biswas, R.: Outlining big data analytics in health sector with special reference to Covid-19. Wirel. Pers. Commun. **124**(3), 2097–2108 (2022)
4. Brändle, T., Colombier, C.: Healthcare expenditure projections up to 2050: ageing and the COVID-19 crisis. (2022)
5. BDVA TF7 Healthcare subgroup: Big Data Technologies in healthcare: needs, opportunities and challenges https://www.bdva.eu/sites/default/files/Big%20Data%20Technologies%20in%20Healthcare.pdf
6. Vaishya, R., Javaid, M., Khan, I.H., Haleem, A.: Artificial Intelligence (AI) applications for COVID-19 pandemic. Diabetes & Metabolic Syndrome: Clinical Research & Reviews. **14**(4), 337–339 (2020)
7. Mehta, N., Shukla, S.: Pandemic analytics: how countries are leveraging big data analytics and artificial intelligence to fight COVID-19? SN Comput. Sci. **3**(1), 54 (2022)
8. Galetsi, P., Katsaliaki, K., Kumar, S.: The medical and societal impact of big data analytics and artificial intelligence applications in combating pandemics: a review focused on Covid-19. Soc. Sci. Med. **301**, 114973 (2022)
9. Chen, J., Chang, S., Zhang, P., Chen, Q., Peng, P., Claramunt, C.: A Critical Examination for Widespread Usage of Shipping Big Data Analytics in China. J. Mar. Sci. Eng. **10**(12), 2009 (2022)
10. https://www.patersonsimons.com/wp-content/uploads/2018/06/TMS_SmartPort_InsightBee_Report-to-GUIDE_01.02.18.pdf
11. Horizon Europe: Healthy Ship 4U (HS4U). https://hs4u.eu/

Developing Housing at Sea: A Case for Humanitarian Assistance and Residency Vessels

Kaitlyn West and Rita Cheung

1 Introduction

According to the World Bank, the term *blue economy* refers to "the sustainable use of ocean resources for economic growth, improved livelihoods, and jobs while preserving the health of the ocean ecosystem" [1]. By this definition, the development of housing at sea in either a residential or humanitarian capacity fits into the goals of a blue economy. Despite economic growth, benefits to improve livelihoods, and providing jobs, there is a distinct lack of scientific information regarding the development of permanent and semipermanent housing, especially humanitarian housing, at sea and its impact on blue growth and development. This chapter describes the current evidence for housing at sea, potential growth strategies, and usages during emergency situations, based on publicly available information and anecdotes.

As countries are facing widespread housing shortages, the prospect of sea-based housing may alleviate the stress on shore-based facilities. With not only opportunities for governments, coastal cities, and other stakeholders, the development of this industry also has the potential to impact individuals through safe, stable housing. Further benefits include job development, reduction in shipbreaking waste, and economic growth through private sales of cruise cabins.

K. West (✉)
COLUMBIA pure, Hamburg, Germany
e-mail: kw.sign@ure.blue

R. Cheung
COLUMBIA blue, Hamburg, Germany
e-mail: rc@columbia.blue

S. Th. Rassia (ed.), *The Blue Book*, https://doi.org/10.1007/978-3-031-48831-3_4

2 Residential Vessel Types and Use Cases

As there is no industry-wide definition for either residency vessels or humanitarian vessels, the below definitions will be used for the purposes of this chapter. *Residency vessels* has been defined as "vessels where cabins are privately and exclusively owned, sold or leased for periods in excess of 30 days out of each calendar year." This is based on the standard leasing and selling terms of current residency vessels (see: Table 1. Residency Vessels by Unit and Price, 2023). Housing at sea, not used for humanitarian purposes, is a relatively new concept, which is under development at many companies globally. In a personal interview with COLUMBIA blue CEO, Norman Schmiedl, COLUMBIA blue and associated partners were first introduced to the concept of residency vessels and humanitarian vessels in 2020 and 2022. He explains that residency vessels are a relatively new and exciting concept for the industry [2]. COLUMBIA blue is an integrated leisure services platform. It aims to set new standards in performance efficiency, state-of-the-art technology, sustainability, and customized service delivery on land, at sea, and in the air. The top management of this company was approached for background information for this article as it has a long history in the industry and have been involved in several projects regarding housing at sea, those these projects remain unnamed to protect client confidentiality.

Humanitarian vessels will be defined, for the purposes of this chapter, as "vessels that have been repurposed for the temporary housing of refugee, homeless and work-force populations during an acute crisis." There is no time frame for the usage of a humanitarian vessel, however, a humanitarian vessel, for the purposes of this chapter, will not include any vessels that are used exclusively for the transport of refugees, rescue of refugees (e.g., Sea-Watch), or other rescue assistance operations. Humanitarian vessels are not a new concept and have been used historically, including during World War II and following Hurricane Katrina, to name a few circumstances.

3 Humanitarian Vessels

Humanitarian vessels act to alleviate the strain on local communities and provide safe housing outside of affected areas. The concept of humanitarian vessels is not a new concept. In fact, in recent history, ships were used for humanitarian purposes during crises like Hurricane Katrina [3], the recent invasion of Ukraine and even during the Second World War. In the USA, one of the first usages of a ship to house displaced persons was in 1989, following the Loma Prieta earthquake in the San Francisco Bay [4]. This section elucidates the definitions of humanitarian vessels, use cases, and high-level considerations required to implement a humanitarian vessel during times of crisis.

Table 1 Residency vessels by publicly available unit and price, 2023

Cruise company	Total units available (estimate)	Administration fees	Small unit size	Small unit price in million euros	Large unit size	Large unit price in million euros	Ownership options
Storylines [37]	547	From $2152 pppm	22 m^2	0.9	133 m^2	8	24 years, or life of the vessel
Blue World [38]	~500	15%	79 m^2	2.47	111.5 m^2	3.52	12 years, 24 years, or life of the vessel
The World [39]	165	10–15%	27 m^2	0.8	300 m^2	7.3	Life of the vessel
Njord [40]	117	5%	124 m^2	8.5	495 m^2	70	Until 2052
Somnio [41]	39	15–20%	182 m^2	9.5	963 m^2	Custom price	Life of the vessel
Clydebuilt [42]	101	Unknown	86 m^2	9.5	587 m^2	130	50 years
Utopia Residences [43]	199 residences, 218 hotel cabins	4.5%	130 m^2	4	613 m^2	36	Life of the vessel

The US Federal Emergency Management Agency (FEMA) has utilized cruise ships to house displaced persons, but also to house those intended to work on disaster management, relief, and mid-term projects in disaster-affected regions [5–7]. FEMA describes cruise ships, recreational vessels, and floating hotels as noncongregate shelter, which provides a higher level of privacy than congregate housing (e.g., stadiums, schools, tent camps, community centers, etc.), which are often used as initial options to house displaced persons at short notice. However, despite their recognition as a short-term solution for housing, they have been used on a medium-term basis in more recent humanitarian crises [3, 5, 8–15]. The feasibility of these projects should always be assessed on a case-by-case basis with community stakeholders and governmental representatives responsible for the assessment.

These humanitarian projects are usually initiated by local or regional governments, which have experienced an acute influx of displaced persons in recent months, and then reach out to brokers to find suitable ships or locales for their projects, whether short- or mid-term. Local hotels, traditionally used as interim housing for displaced persons and during humanitarian crises, are not always an ideal mid-term solution for many reasons.

First and foremost, the cost of housing displaced persons and asylum seekers can be astronomical to governments and taxpayers. In a UK House of Commons Home Affairs Committee meeting on the 26 October, 2022, it was publicly announced that the average cost of housing displaced persons and asylum seekers in hotels amounted to around 6.8 million GBP per day and that the cost was likely to increase [16]. The UK Government has been quoted saying that vessels, as opposed to hotels, provide a cost effective and sustainable accommodation [17]. More research regarding the financial durability of humanitarian vessels is needed to confirm the overall cost-effectiveness of these programs. Seeking alternative solutions, such as ships, can significantly reduce the cost burden, however, the exact costs are not publicly available. Norman Schmiedl of COLUMBIA blue confirms this, stating that utilizing ships for humanitarian needs can be cost effective and highly beneficial due to the resources available on board, compared to those available in a hotel [2].

In some cases, when utilizing hotel accommodation, the displaced persons are left without access to critical resources and support structures that can assist in integration (e.g., transportation, health care access, etc.). In addition, it is possible that the social support workers (i.e., case workers and social services) are left spread thin across the city with individuals and families in numerous hotels and facilities across the region. In October 2022, the Irish Times published an article expressing concern for the frontline workers managing displaced persons. The article specifically sights concern with locating proper accommodation as a time consuming and complex for case workers [18].

Furthermore, strain on hotels during acute refugee crises has also been documented. According to S. Ivanov, T.A. Stavrinoudis, in an article published following geopolitical instability in the Middle East in 2015, the consequences for affected hotels on several Greek Islands were highly negative [19]. The article describes negative impacts such as cancellation of hotel reservations, flights, cruise ship calls; loss of income; and decrease in overall bookings during the crisis. The research further investigated the impacts by surveying hotel managers, which were affected by the crisis. The results showed that all operational statistics deteriorated, cost cutting measures were implemented (rather than operating with less employees), marketing outreach was increased, and supplier payments were delayed accommodating the decrease in tourist bookings. The potential consequences for the hotel industry and tourism industry are clear, however these could be mitigated by utilizing alternative housing solutions, such as vessels, during acute crises.

Mitigation measures to decrease pressure on local resources vary globally, however many countries in the European Union have dispersal policies, which require displaced persons to be spread across various localities to reduce the burden on the regional authorities and improve access to social services. However, despite such dispersal policies, some countries still struggle to keep up with sudden influxes of displaced persons during times of crisis. Moreover, according to the European Social Network, even countries with such policies have local and regional solutions that cannot always provide the same level of support to displaced persons [20]. In these cases, access to health care, childcare, and poor sanitary conditions may become issues during the project period. Utilizing ships can help to eliminate some of these issues by providing integrated services on board.

Passenger vessels used for humanitarian projects may provide a suitable solution to the problems experienced by countries with and without dispersal policies due to their mobility and resources available on board. Passenger vessels not only provide a mobile environment that can be repositioned if needed due to strain on local resources, but they also contain most equipment to fully support the population on board, to a reasonable extent. For example, nurses and doctors can be employed directly on board to provide medical services, food can be cooked and prepared directly on board, childcare can be provided directly on board, and sanitation must adhere to the international and local standards of passenger vessels, ensuring a high standard of sanitation – in the case of Europe, the EU SHIPSAN Act [21], and in the case of the USA, the US Vessel Sanitation Program [22]. The options to address care needs can be tailored to each community when utilizing passenger vessels for humanitarian needs.

When determining suitability, both the vessel and location must be assessed. To begin the process, communities must ensure that ships are available for usage in their areas, that berths are large enough to accommodate the ship, and that the port is in an area with resources required (e.g., grocery stores, public transit, etc.). In addition, there must be enough resources to provide the required services either directly on the ship or as close as practicable.

4 Modern Humanitarian Vessel Projects

Recently, humanitarian vessel projects were underway in both the UK and the Netherlands, while considerations were given to housing displaced persons on vessels in the USA. All the modern projects below are elaborated based on publicly available information. At this time, no scientifically published information was available regarding the usage of vessels for humanitarian purposes in a modern context.

In the UK, the Scottish Government has been proactive in the utilization of vessels for Ukrainian displaced persons seeking assistance. As of June 2022, following the invasion of Ukraine by Russia, the UK received 190,537 visa applications from Ukrainians, an increase of 17% from the previous year when just 33,862 applications were received [23]. The extreme need for quick and reliable housing solutions was apparent. Two vessels were secured near Glasgow and Edinburgh, Scotland, in 2022, to house a large influx of people at the time. The ships and their locations were developed with access in mind. The ships had a 24-h shuttle bus to take the passengers to the city and support services, including health care and benefits support located directly on board [14] As of early 2023, the Scottish Government was working to secure long-term housing solutions for those displaced persons on board with as little disruption as possible [24]. The UK has now also expanded their provision of vessel-based housing for humanitarian needs, as of April 2023. The government will now utilize a barge in Portland Port to house 500 single adult males with health care, catering, and security all offered on board. It has been cited as a cost-effective solution to housing unexpected migrant fluctuations. The vessel will remain for at least 18 months to support the needs of migrants entering the country undocumented on small boats crossing the channel [17].

In the USA, the New York State seriously considered utilizing cruise ships to manage the influx of displaced persons from Central and Southern America in late 2022. Asylum seekers were instead placed in 23 temporary shelters established across the city and one cruise ship terminal, which could only accommodate men and did not provide private accommodation [9, 25]. This project was not the first attempt in the USA to house people experiencing homelessness. In 2019, after the number of people experiencing homelessness increased by 47% over the course of 2 years, the city of Oakland considered utilizing cruise ships to temporarily house approximately 1000 people experiencing homelessness [26]. These projects are still under investigation and not currently active.

In Rotterdam, a Holland America cruise ship was repurposed to house over 1300 displaced persons in 2022. The alternative for the displaced persons in the Netherlands was to visit a notoriously dangerous and disconnected displaced persons housing project ashore, which had to place temporary beds on the floors to accommodate all the asylum seekers in 2022. As a solution to the crisis, the city of Rotterdam established the humanitarian vessel in the city center to house Ukrainian displaced persons, alleviating the pressure on shore-based facilities to make space for additional residents [27].

5 Residency Vessels

Private residence at sea is a luxury typically reserved for affluent travelers. However, given the wide range of pricing and size options on residency cruise vessels, this opportunity is not always necessarily reserved for the ultra-wealthy. Not only are there several companies designing residency vessels for seafaring travelers, but also the concept of affordable housing at sea is being explored by cities looking for alternatives to building new housing and infrastructure ashore, though this concept seems to be at an early stage of development.

At the time this chapter was written, only a few companies were developing residency vessels with the options to purchase or partially purchase a residential unit on board a ship. A comparison of the several companies with residency vessel concepts has been described in Table 1 (see below). At present, several residency vessel companies are targeting luxury and ultra-wealthy members for their rooms, though some offer more affordable options. As seen from the examples in Table 1, the price to be a full or partial owner of a cruise cabin ranges from under a million to multiple million euros, which indicates a range of availability and price points suited to wealth and ultra-wealthy patrons.

The implementation of these projects at such an expensive level begs the questions, "what if sea-based housing could be made even more affordable?" and "is there an opportunity to supplement the local housing markets with vessel-based housing?". These questions were recently asked by architecture firms after the reduction in global cruise ship tonnage following the early COVID-19 pandemic phases when passenger ships were decommissioned and sold at one of the highest rates in history [28].

The architects proposed solutions specific to Miami, though these could be implemented in numerous other localities, as there is also a growing lack of affordable housing in the area. As part of the research, residents of Miami were surveyed about whether they would be interested in living in a "coastal housing scenario." According to the survey, a significant proportion of respondents would be interested in affordable housing at sea [28]. The architects did not complete the project, however, as it was only conceptual and established for the purposes of research only.

Other media outlets report a few other projects that are looking at the possibility of using ships as housing. In Portland, Maine in the USA, the option of using cruise ships as transitional housing for those experiencing homeless and low-income families was investigated in 2018 but the project was abandoned due to criticisms. At present there are very few publicly announced low-income or affordable projects in this field. However, as climate change begins to affect coastal communities, it is recommended that coastal cities consider ship-based housing as an alternative.

6 Blue Economy Benefits to Housing at Sea

There are several assumed blue economy benefits to providing housing at sea, both privately and as a humanitarian solution. From reduction of environmental waste generated by shipbreaking to infrastructure and workforce development, the benefits to coastal cities and the environment are evident.

During times of market volatility and variability, utilizing old, or otherwise unused tonnage for humanitarian or residency vessels allows companies to ensure minimal loss and support the blue economy. For example, following the COVID-19 pandemic, the global cruise market shrank significantly. In 2019, there were approximately 27.5 million passengers carried on cruise ships globally, while by 2020 (during the COVID-19 pandemic) that number had decreased to just 7.1 million. The number of cruise ship passengers has not yet recovered to reach pre-pandemic levels, with just 13.9 million passengers in 2021 [29]. By utilizing ships as alternative housing, ship owners can provide useful solutions to coastal communities with humanitarian or housing needs, as well as provide themselves with a source of revenue in unsure times.

The usage of vessels for residential housing or humanitarian purposes can also contribute to limiting waste and environmental risks resulting from shipbreaking practices. It has been shown that the toxic metal leakage from shipbreaking sites significantly affects the water and sediments in the surrounding areas [30, 31]. In 2019, Bangladesh, India, and Pakistan were the primary countries for shipbreaking. There, several areas have been reported as moderately or highly polluted due to toxic metal contamination [30, 32]. Utilizing ships as residences or humanitarian vessels offers a more environmentally sustainable alternative to ship disposal, especially during periods of crisis in which it is not possible for ships to operate as originally intended.

Furthermore, the usage of vessels as permeant or semipermanent residences can often have unexpected consequences on the infrastructure a shore. In fact, humanitarian and residency vessels often have more ideal proximity to shore-based resources compared to camps or repurposed old buildings. The presence of residential or humanitarian vessels can lead to a change in the local terminal operations and activities. This is due to the variety of resources such vessels require to be fully operational and to provide all services to the residents. Overall, port services are most likely to undergo development and expansion to accommodate the needs of those permanently or semipermanently residing on vessels. Moreover, the presence of ship residents determines the need for increased transportation solutions that connect the port to neighboring towns and cities. Within this context, coastal cities have the opportunity to further develop their public transportation infrastructure and increase the accessibility of the port. All these changes are deemed to benefit the development of the blue economy by preserving livelihoods and ensuring social inclusion [33].

Lastly, workforce development is another key blue growth benefit deriving from the usage of residency and humanitarian vessels. Especially regarding humanitarian vessels, where possible, the local and on-board populations can often be hired to work on board (depending on visa and seafarer qualification requirements), gaining skills that can be later applied ashore once permanently settled.

7 Operational Constraints and Considerations

Operators must consider whether the risks and rewards of a project focused on housing at sea are for the benefit of all parties involved. Especially when it comes to humanitarian housing, homeless housing, and other humanitarian relief efforts, these projects are often public and heavily criticized as these populations require more intensive care and resource allocation than those wishing to live on a vessel for leisure or recreation.

In addition, regarding housing at sea and residency vessels, operators and owners must consider several aspects and ensure feasibility analyses are conducted prior to implementation. Aspects may include location, flag state requirements, workforce visas, and local requirements for housing as these may all impact the operation, especially those with more permeant or stationary localities.

The example of the Satoshi can be used as a tale of caution when trying to establish a residency vessel. The Satoshi was a cryptocurrency-based residency vessel, which aimed to establish a floating community of the coast of Panama. While the vessel and the company, Viva Vivas, had the support of the Panamanian government, the project fell short due to lack of market research and preparedness by the owners. Tasks like registration, compliance with international safety documentation, maintaining the vessel as a vessel rather than a hotel, and waste discharge, to name a few, were ultimately responsible for the project's collapse [34]. However, with a well-formulated business plan and proactive research on the intricacies of the industry, a successful residency program is possible. It is key that all projects in this field are well researched and backed by the local communities to ensure success.

Another potential operational constraint is the affected population, which the vessel is intended for. Stakeholders must consider if the affected population would be willing to stay on board the vessels given the circumstances of any humanitarian or local issues. In the humanitarian crisis that followed hurricane Katrina in New Orleans, Louisiana, USA, in 2006, which flooded almost 80% of the City of New Orleans, led to more than 1800 deaths, and caused more than 160 billion US dollars in damage in New Orleans [11], three cruise ships were chartered by FEMA.

One of the criticisms of the FEMA contract was that recent survivors of flood waters and severe weather did not wish to be on a ship in the very waters that had taken their homes, family members, and livelihoods [6]. Additionally, after hurricane Ida, FEMA paid high prices for cruise ships that were ultimately not needed. Many cited that the electricity was restored in New Orleans much faster than anticipated [35]. This illustrates the need for a balance between the needs of a community, the anticipated usage, and the willingness to use a ship. Communities, which would be willing to use cruise ships in an emergency setting should be identified and ships added to their emergency planning where possible.

Additional operational constraints and considerations include the potential for highly volatile markets, though this can also be an advantage in times of great need. For instance, the COVID-19 pandemic led to a significant decrease in the number of cruise ship passengers worldwide in 2020, leading to a crisis in the global cruise market. Despite a recovery in 2021, the industry has yet to reach its pre-pandemic performance in terms of passenger flow [29]. However, the loss to companies could be balanced by utilizing ships, which have been out of service due to these disruptions.

Then again in 2022, the Russian invasion of Ukraine caused major disruptions to supply chains, workforces, and importations. It was estimated that Russia and Ukraine provided up to 14.5% of the total global shipping workforces, making it difficult to procure qualified crew members in the industry [36]. Without these workforce members, the cost of recruiting and training new seafarers has risen, leading to further unexpected costs and crisis for the industry. However, as discussed above, in times of crisis the affected populations can be recruited to work on board the ships, particularly in humanitarian situations. This can be a reliable source of labor for companies and further help in the development of workforces within coastal communities where displaced persons may reside.

8 Limitations

Limitations to research in this field include a lack of consistent and publicly available evidence. Due to the highly competitive nature of the hospitality industry, much information regarding operating costs and expenditure is kept confidential and unavailable for the purposes of research. However, initiatives such as Healthy Ship 4 U and other similar governmentally funded programs aim to unite the scientific community with modern operators to enhance the scientific understanding of life on board cruise ships. It would be highly beneficial to future research to seek out and participate in projects with international and regional funding sources to ensure transparency and efficacy of humanitarian and social initiatives such as residential vessels.

9 Conclusion

In conclusion, there are numerous options to utilize vessels as alternative housing solutions to combat housing shortages on land and to expand blue growth and bring people to the seas in a meaningful way. From reduction of shipbreaking waste to workforce and infrastructure development, the benefits of humanitarian and residential housing are evident. In addition, the benefits to humans are also clear including coastal access in cities, potential for low-income housing solutions and safe housing during acute crises.

Utilizing ships in alternative ways not only helps the blue economy develop, but it also reduces strain on shore-based resources and provides a safe and healthy environment for people to thrive. This chapter intends to provide a platform for ship owners, governments, and other stakeholders on which to model their alternative housing options, especially when coastal access is readily available. While ships require some additional resources to properly operate due to their technical requirements, they can ultimately provide a place for people to thrive alongside our oceans, if planned appropriately.

Further research is needed to ensure that effective solutions are in place in alignment with the goals of blue growth and development. Details such as financial data, satisfaction (especially regarding humanitarian projects), and outcomes are not currently published. In addition, when it comes to residency vessels, more information will become available upon the successful launch of these projects.

References

1. What is the Blue Economy? [Internet]. World Bank. [cited 2023 Feb 6]. Available from: https://www.worldbank.org/en/news/infographic/2017/06/06/blue-economy
2. Schmiedl, N.: Blue Economy and Housing. (2023)
3. NBC News. Cruise ships chartered for hurricane refugees. NBC News [Internet]. 2005 Sep 3 [cited 2023 Feb 6]. Available from: https://www.nbcnews.com/id/wbna9188316
4. Romero, D.: Housing crisis spurs some to consider old cruise ships as solution to house needy. NBC News [Internet]. 2018 Oct 7 [cited 2023 Apr 18]. Available from: https://www.nbcnews.com/news/us-news/housing-crisis-spurs-some-consider-old-cruise-ships-solution-house-n917301
5. Carnival Cruise Lines. Statement regarding carnival fascination. Carnival Cruise Line News [Internet]. 2017 [cited 2023 Feb 3]. Available from: https://carnival-news.com/2017/10/12/statement-regarding-carnival-fascination/
6. Cruise, W.L.: Ships, spurned by evacuees, are home to displaced city workers and families. The New York Times [Internet]. 2005 Sep 16 [cited 2023 Jan 17]. Available from: https://www.nytimes.com/2005/09/16/us/nationalspecial/cruise-ships-spurned-by-evacuees-are-home-to-displaced.html
7. FEMA paid millions for half-empty "floating hotel" after hurricane Maria [Internet]. WLRN. 2018 [cited 2023 Jan 11]. Available from: https://www.wlrn.org/news/2018-05-30/fema-paid-millions-for-half-empty-floating-hotel-after-hurricane-maria
8. Disused cruise ships could be used to house asylum seekers – Braverman. BBC News [Internet]. 2022 Dec 21 [cited 2023 Jan 10]. Available from: https://www.bbc.com/news/uk-politics-64053099
9. Fitzsimmons, E.G., Newman, A.. Eric Adams considers housing migrants on cruise ships. The New York Times [Internet]. 2022 Sep 19 [cited 2023 Apr 19]. Available from: https://www.nytimes.com/2022/09/19/nyregion/eric-adams-migrants-cruise-ships.html
10. Glaze, B.: Housing migrants on cruise ships "on the table" as system to cost £3.5bn a year [Internet]. Mirror. 2022 [cited 2023 Jan 11]. Available from: https://www.mirror.co.uk/news/politics/housing-channel-migrants-cruise-ships-28786959
11. Encyclopedia Britannica: Hurricane Katrina – Aftermath [Internet]. Encyclopedia Britannica. 2023 [cited 2023 Feb 6]. Available from: https://www.britannica.com/event/Hurricane-Katrina

12. Vlamis K.: Over 2,000 Ukrainians have spent the past 5 months living on cruise ships off the coast of Scotland: "it was like a small social experiment" [Internet]. Business Insider. [cited 2023 Apr 19]. Available from: https://www.businessinsider.com/ukrainians-living-on-cruise-ship-in-scotland-for-5-months-2023-1
13. Scottish Government: Second charter ship for Ukrainians [Internet]. 2022 [cited 2022 Dec 16]. Available from: http://www.gov.scot/news/second-charter-ship-for-ukrainians/
14. BBC: Second cruise ship for Ukrainian refugees arrives in Glasgow. BBC News [Internet]. 2022 Jan 9 [cited 2023 Feb 1]. Available from: https://www.bbc.com/news/uk-scotland-62757777
15. Cohen P.: When home is a ferry ship: an influx from Ukraine strains Europe. The New York Times [Internet]. 2022 Aug 4 [cited 2022 Dec 16]. Available from: https://www.nytimes.com/2022/08/04/business/economy/ukraine-refugees-estonia.html
16. Channel crossings, HC 822 [Internet]. House of Commons; 2022 [cited 2023 Jan 10]. p. 37. Available from: https://committees.parliament.uk/oralevidence/11390/pdf/
17. Vessel to accommodate migrants [Internet]. GOV.UK. [cited 2023 Apr 19]. Available from: https://www.gov.uk/government/news/vessel-to-accommodate-migrants
18. The Irish Times: Frontline workers fear consequences of stretching refugee accommodation too thin. The Irish Times [Internet]. 2022 Oct 22 [cited 2023 Feb 2]. Available from: https://www.irishtimes.com/ireland/2022/10/22/ngo-workers-fear-consequences-of-stretching-refugee-accommodation-too-thin/
19. Ivanov, S., Stavrinoudis, T.A.: Impacts of the refugee crisis on the hotel industry: evidence from four Greek islands. Tour. Manag. **1**(67), 214–223 (2018)
20. Montero A.L., Baltrucks, D.: The impact of the refugee crisis on local public social services in Europe [Internet]. The European Social Network; 2016 Aug [cited 2023 Apr 17]. Available from: https://ec.europa.eu/migrant-integration/library-document/impact-refugee-crisis-local-public-social-services-europe_en
21. EU SHIPSAN. EU SHIPSAN ACT Joint Action [Internet]. 2023 [cited 2023 Apr 17]. Available from: https://www.shipsan.eu/
22. National Center for Environmental Health. Vessel Sanitation Program | VSP | CDC [Internet]. 2023 [cited 2023 Apr 17]. Available from: https://www.cdc.gov/nceh/vsp/default.htm
23. Statistics on Ukrainians in the UK [Internet]. GOV.UK. [cited 2023 Apr 19]. Available from: https://www.gov.uk/government/statistics/immigration-statistics-year-ending-june-2022/statistics-on-ukrainians-in-the-uk
24. Ferguson L., Clark J.P.: Ukrainians living on ship in Glasgow set to be rehomed across Scotland [Internet]. Daily Record. 2023 [cited 2023 Apr 19]. Available from: https://www.dailyrecord.co.uk/news/scottish-news/ukrainians-living-ship-glasgow-set-29481979
25. de Freytas-Tamura, K., Zraick, K.: New York City to open migrant shelter at Brooklyn Cruise Terminal. The New York Times [Internet]. 2023 Jan 21 [cited 2023 Apr 19]. Available from: https://www.nytimes.com/2023/01/21/nyregion/migrant-homeless-cruise-terminal.html
26. USA Today. California official floats a new idea: House homeless on a cruise ship [Internet]. USA TODAY. [cited 2023 Apr 24]. Available from: https://www.usatoday.com/story/news/politics/2019/12/11/oakland-councilwoman-proposes-housing-homeless-cruise-ship/4400158002/
27. Chan, W.: What happens when you house more than 1,000 refugees on a cruise ship? [Internet]. Curbed. 2022 [cited 2023 Feb 17]. Available from: https://www.curbed.com/2022/11/rotterdam-ukrainian-refugees-cruise-ship-holland-america.html
28. Tayeb, Z.: How decommissioned cruise ships could be given a new lease of life as affordable housing for Miami residents [Internet]. Business Insider. [cited 2023 Feb 6]. Available from: https://www.businessinsider.com/cruise-ships-repurposing-decomissioned-affordable-housing-miami-florida-2022-2
29. Cruise Market Watch: Worldwide passengers carried [Internet]. Tableau Software. 2022 [cited 2023 Apr 24]. Available from: https://public.tableau.com/views/2012Growth/2021Growth?%3Adisplay_static_image=y&%3AbootstrapWhenNotified=true&%3Aembed=true&%3Alanguage=en-US&:embed=y&:showVizHome=n&:apiID=host0#navType=0&navSrc=Parse

30. Ali, M.M., Islam, M.S., Islam, A.R.M.T., Bhuyan, M.S., Ahmed, A.S.S., Rahman, M.Z., et al.: Toxic metal pollution and ecological risk assessment in water and sediment at ship breaking sites in the Bay of Bengal Coast, Bangladesh. Mar. Pollut. Bull. **175**, 113274 (2022)
31. Reddy, M.S., Basha, S., Sravan Kumar, V.G., Joshi, H.V., Ramachandraiah, G.: Distribution, enrichment and accumulation of heavy metals in coastal sediments of Alang–Sosiya ship scrapping yard, India. Mar. Pollut. Bull. **48**(11–12), 1055–1059 (2004)
32. Chapman, K.: The toxic tide of ship breaking [Internet]. Chem. World. 2022 [cited 2023 May 4]. Available from: https://www.chemistryworld.com/features/the-toxic-tide-of-ship-breaking/4015158.article
33. United Nations: Blue Economy: oceans as the next great economic frontier [Internet]. United Nations Western Europe. 2022 [cited 2023 May 4]. Available from: https://unric.org/en/blue-economy-oceans-as-the-next-great-economic-frontier/
34. Elmhirst, S.: The disastrous voyage of Satoshi, the world's first cryptocurrency cruise ship. The Guardian [Internet]. 2021 Sep 7 [cited 2023 Apr 24]. Available from: https://www.theguardian.com/news/2021/sep/07/disastrous-voyage-satoshi-cryptocurrency-cruise-ship-seassteading
35. Karlin S. After Ida, Carnival cruise ship housed 103 key workers, costing FEMA nearly $50K per head. NOLA.com [Internet]. 2021 Sep 21 [cited 2023 Feb 28]. Available from: https://www.nola.com/news/after-ida-carnival-cruise-ship-housed-103-key-workers-costing-fema-nearly-50k-per-head/article_5d1123c6-1b1c-11ec-b407-c31b869f71db.html
36. Consultancy.eu. How the Russia-Ukraine conflict is impacting supply chains [Internet]. 2022 [cited 2023 Apr 24]. Available from: https://www.consultancy.eu/news/7993/how-the-russia-ukraine-conflict-is-impacting-supply-chains
37. Storylines. Available Residences [Internet]. 2022 [cited 2023 Jan 30]. Available from: https://www.storylines.com/listings-5
38. Blue World Voyages, LLC: Blue World Residences – unpack once & see the world [Internet]. Blue World Residences. [cited 2023 Jan 30]. Available from: https://www.blueworldresidences.com
39. The World. The World – Luxury Residences at Sea [Internet]. The World Residences at Sea. 2022 [cited 2023 Mar 1]. Available from: https://aboardtheworld.com/
40. Ocean Development Group. Superyacht NJORD – Ocean Residences [Internet]. [cited 2023 Mar 1]. Available from: https://thefinestaddress.com/
41. Somnio Superyachts [Internet]. Somnio Superyachts. 2022 [cited 2023 Mar 1]. Available from: https://somniosuperyachts.com
42. Clydebuilt. Specification | Clydebuilt [Internet]. 2021 [cited 2023 Mar 1]. Available from: https://www.clydebuilt.com/specification/
43. Utopia Residences | Residences [Internet]. [cited 2023 Mar 1]. Available from: http://www.utopiaresidences.com/residences/index.php

Smart Coastal City Mobility Management in the Context of High Tourist Activity and Methodological Approaches

Bertrand David, Chuantao Yin, and René Chalon

1 Introduction

The Smart city originated in the media sector, which refers to integration of urban systems and services by using various information technologies or innovative concepts to improve the efficiency of resource utilization, optimize urban management, and improve the quality of life of citizens.

Today, countries worldwide realize in varying degrees the idea of smart cities and the different implementation efforts. However, there is no denying that in the twenty-first century, the age of smart is here. The Smart City is not only the application of new information technologies, but also concerns the participation of citizens in the various activities of the city with the intelligence of humans, combined with Artificial Intelligence in different forms. A variety of classifications try to clarify this huge domain. One of these proposes six dimensions: environment, economy, living, mobility, people, and government with the following orientations: wireless city, smart home, smart transportation, smart public service and social management, smart urban management, smart medical treatment, green city, and smart tourism [1].

It is also possible to examine whom these evolutions concern. By avoiding the generic "to everybody," we classify the answers by age: kids, teenagers, adults, seniors; by implication in the city: citizen, neighbor, student, worker, administrator, governor; without avoiding important orientation to all persons with deficiencies.

B. David (✉) · R. Chalon
Université de Lyon, CNRS, Ecole Centrale de Lyon, LIRIS, UMR5205, Lyon, France
e-mail: Bertrand.David@ec-lyon.fr; Rene.Chalon@ec-lyon.fr

C. Yin
Sino-French Engineer School, Beihang University, Beijing, China
e-mail: chuantao.yin@buaa.edu.cn

S. Th. Rassia (ed.), *The Blue Book*, https://doi.org/10.1007/978-3-031-48831-3_5

We can also try to draw up a list of reasons for: better common welfare and neighborhood, energy, transportation of goods and passengers, information dissemination on culture, sports, social services, tourism, etc.

The Smart City is an area where all ICT possibilities can be used: data accessibility, data processing, information access exchange and manipulation in static and mobility situations, using wired or wireless networks, mainly in contextual situations. Multiple ICT technologies are used to solve these problems such as Internet, Internet of Things (IoT), Location-Based Services (LBS), Big Data and Open Data, as well as Artificial Intelligence with Machine and Deep Learning.

This chapter is organized in the following manner: After a short introduction and a state of the art, we present a conceptual view with assistive system role, functions, and supports. Data identification and dynamicity management are important aspects, as well as architectural physical and digital infrastructures, open data and artificial intelligent treatments. In the next section we present a design methodology with five identified approaches and we describe their use in different cases. At the end, we shortly present appropriate smart city services for disable persons and we conclude giving our expected orientations.

2 State of the Art

To carry out a complete state of the art for the Smart City would appear to be totally impossible as the amount and diversity of work is large and diversified. It would be hard enough to establish a common definition and classification, and indeed this has not yet been achieved, while listing all approaches and contributions seems totally impossible. We thus propose that the reader consults our paper [2]. In relation to transportation aspects, a relevant reference is [3]. The Japanese view seems interesting [4], as is also Singapore's perception of emerging technologies [5]. The questions on the effectiveness of smart cities make for instructive reading [6], as does also the view presented to CIOs [7]. [8] directly relates to our approach. We can split it into two viewpoints: the users' one and the technological one.

From the users' viewpoint, the list of applications is unbounded, as we mentioned earlier. From the technological one, i.e., from the implementation one, the list is also over abounded, but shorter to explain: software architecture, Internet, Cloud computing, Internet of Things (IoT), Location-Based Services (LBS), Open Data (OD), Machine and Deep Learning (ML & DL), and Human Computer Interaction with Virtual and Augmented Reality and Metaverse.

From the coastal city point of view, we shall now suggest three references [9–11].

3 Coastal City Characterization and How It Becomes Smart

To show how a city evolves into a smart city, specifically in the case of a coastal city, we need first to characterize what a Coastal city is [25]. Then, we identify progressively different dimensions of its evolution into a Smart City. We limit our work to the dimension of management of mobility and the assistive system, the role of which is to manage corresponding behaviors (Fig. 1).

We start with the definition of a Coastal City: A Coastal City is a city that is located by the sea or the ocean. Its activities are related to the sea, i.e., fishing, boat transportation, boat and ship repairs, as well as possibly boat and ship construction. A large majority of coastal cities in France also have important tourist activities related to the proximity of more or less popular beaches. Swimming, diving, snorkeling, yachting, kitesurfing, and fishing are common practices.

This aspect of welcoming tourists is dominant in the organization of City life. It introduces the notion of the high and low seasons, mainly by the number of inhabitants that can vary considerably between these seasons (their number can be multiplied by 10,100, or more in summer). As such, the requirements of life organization are totally different. In the low season, a few inhabitants can live quietly without stress, noise, or traffic jams. However, economically speaking, this period is a "waiting" period for a large majority of inhabitants who rely on their tourist businesses to live (retired people excluded), whether in the form of restaurants, shops, accommodation, or recreation activities.

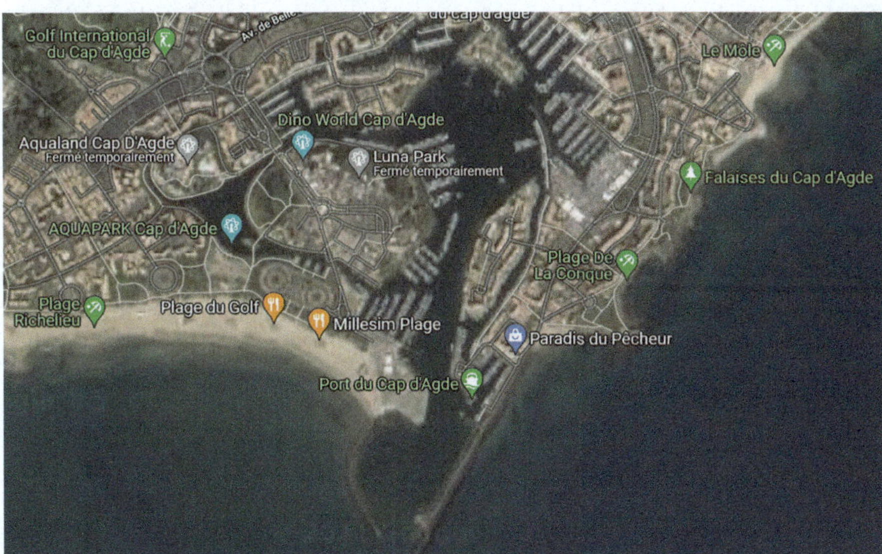

Fig. 1 Coastal city with downtown, harbor, port, beaches, and boating industrial zone

It seems natural to study the evolution of this kind of city into a smart city in the context of high season city life, while only later showing that in the low season it is possible to propose appropriate limited behaviors and city services adapted to this season and its inhabitants.

We begin our study with the general problem of circulation in a city with a large population density. We can inherit classical smart city solutions for different kinds of mobility: pedestrian, bike, scooter, motorbike, car, truck, and public transportation (bus and, for large cities, tram, and subway). The solutions implemented are traffic reorganization with pedestrian zones, one-way streets and roads, prioritized lanes for public transportation, emergency services, and taxis, pedestrian, and bike lanes. In coastal cities two evolutions can be considered: (1) organization of circulation between specific zones, i.e., park-and-ride and downtown, downtown and recreation areas and beaches, etc. and (2) new transportation modes such as boat taxis and individual boats, pedal-boats. New protected lanes, on roads and on the water, can also be created for automated shuttles between these zones.

It is also important to consider the dynamicity of these reorganizations, which can be permanent (available all the time) or evolve according to the time scale (each month, week, day, weekdays, weekend, morning, afternoon, evening, night, per hour, or in unpredictable exceptions). As we will see later, signaling must be appropriate (static or with appropriate dynamicity), as must also be propagation of applicable rules to different users. The goal is evolution of the circulation map during the high season in order to optimize fluidity and preferential transportation. Creation of park-and-ride can be associated with public automated transport to the city, the harbor (port), and the beaches. Management of priority for several kinds of vehicles such as buses, delivery trucks, emergency vehicles, and vehicles for disabled people can also be considered.

3.1 Management of Circulation in the City, in the Harbor/Port, on the Sea

Let us now recall an approach to mobility devoted to dynamic lane management on which we worked [12]. Its objective was to create these lanes dynamically only when buses are present and to leave all lanes open to general traffic when buses are not present. This allows management of general traffic speed. The main technologies used are: a Location-Based Service integrating bus detection sensors; an intermediation platform collecting sensor information and determining dynamic bus section activation and deactivation; in-the-field infrastructure; and/or embedded vehicle interface receiving instantaneous information on selected situations (Fig. 2). From the HCI point of view, it is important to indicate the present situation on in-the-field indicators, as well as on the screen in the vehicle.

Another non-coastal specific service is devoted to better and easier parcel delivery to inhabitants or stores. In this case, we proposed to create delivery areas,

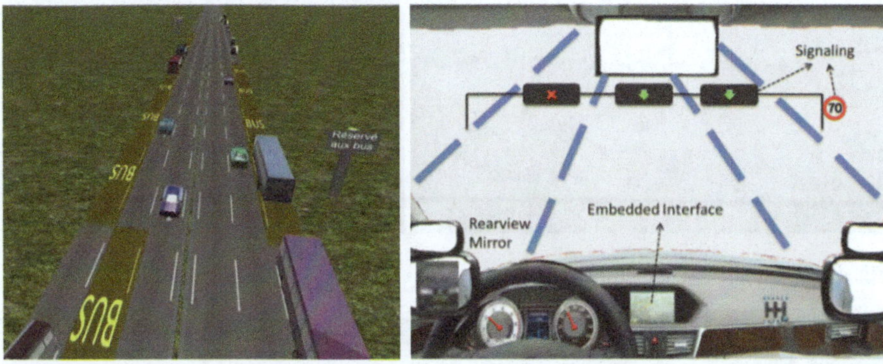

Fig. 2 Dynamic lane management [12]

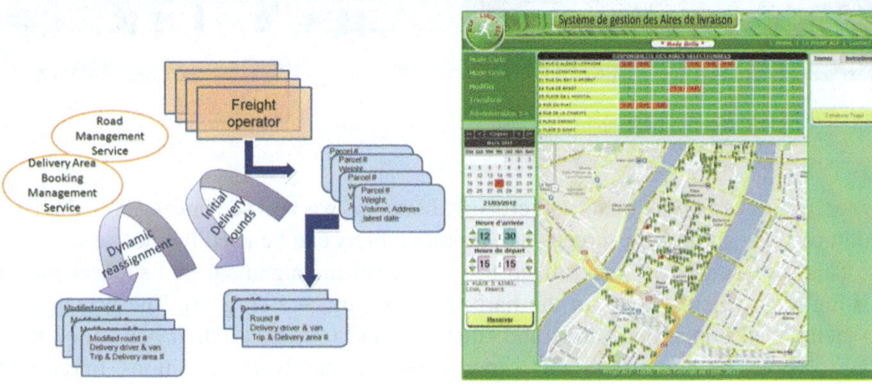

Fig. 3 Parcel distribution based on delivery area reservation [12]

which can be reserved as priority by delivery vehicles, in order to shorten delivery times [13]. In this context, the main data are: delivery addresses of parcels to be distributed and the geographical location of delivery areas to be reserved. An algorithm or interactive tool is used to create a delivery trip integrating the schedule of use delivery areas allowing transported parcels to be delivered to destination addresses. A mobile interaction tablet-based tool allows the driver to modify the trip if a traffic jam problem or other problems occur (Fig. 3).

Management of circulation in the commercial port/harbor is totally specific and has no relation to the touristic part of the city, in order to preserve appropriate working conditions in this part of the city and on the sea (Fig. 4).

In the context of the coastal city, a specific service is devoted to *management of circulation in the port,* taking into account weather conditions and the state of the sea, as well as access to different parts of the port (industrial, professional, sailing, and yachting). In this context, not only local mobility can be managed, but also all

Fig. 4 Industrial coast and navigation zones

Fig. 5 Cruising and fishing boats

the activities of fishing boats and cruising boats (as we will see later). From the HCI point of view, integrated screens and mobile tablets can be used by the crew.

This service can be generalized to the overall management and supervision of fishing and cruising boats in the harbor and on the sea (Fig. 5). The principle is to identify all concerned actors and situations and to indicate their behaviors, their interrelations, and contributions to appropriate trip and work conditions. In the past we studied this kind of ecosystem for buses (e-bus) and trucks (e-truck) [14]. We adapted them to these situations (fish boat and cruising management ecosystems) as schematized in Figs. 6 and 7.

3.2 *Beaches and Sea*

Another important aspect for coastal cities is management of mobility on the beaches and on the sea. On the beach several activities are possible, more or less dynamic: sunbathing, walking along the seashore, carrying out various sports or family activities. In all cases, it is important to look out for other beach users, to avoid accidents and ensure do not disturb them. In the recent COVID period, on many French beaches, city mayors at local level or, more globally, government authorities have determined several categories of rules: prohibit access to beaches, authorize only "dynamic use of the beach" (no sitting or sunbathing, only walking and running), respect of distances between groups of users, either natural or physically materialized zones, with the possibility of reserving them (Figs. 8 and 9).

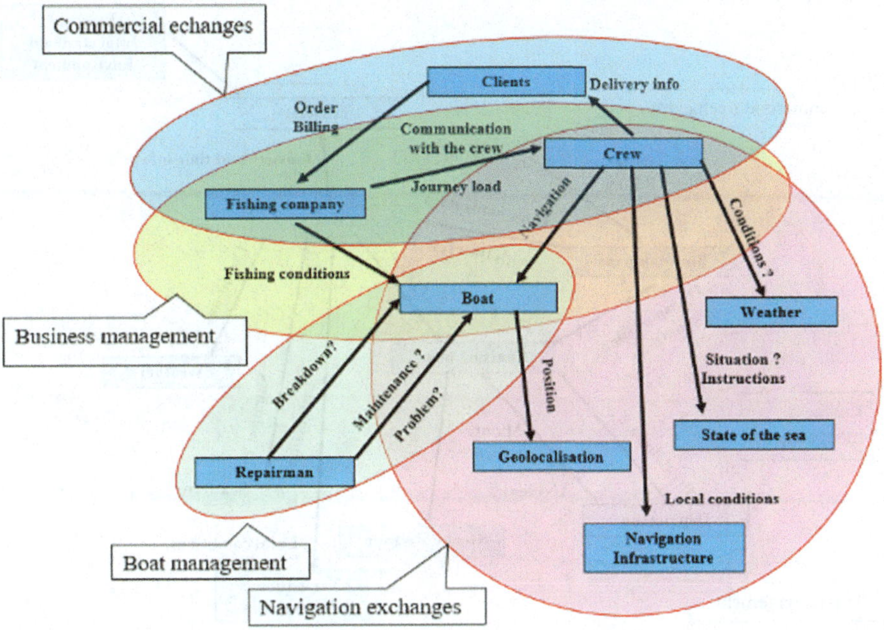

Fig. 6 Fishing boat information ecosystem

On the sea (Fig. 10), the creation of activity zones is a common approach resulting in the creation of specific zones with supervised swimming, diving, and snorkeling activities, as well as sailing, yachting, surfing, kitesurfing, with either visual supervision, or more instrumented electronic monitoring. In this case, each actor (swimmer, sailor, kite surfer) has at his/her disposal a connected watch, or more sophisticated devices able to communicate his/her position and related situation (OK, emergency call, etc.). On the beach, life guards are equipped with a receiver allowing them to monitor the situation and trigger the appropriate rescue actions.

Another service that can be used in downtown, near the parking area, on the harbor and near the beaches, is an information shelter [15]. Its objective is to provide potential passengers and visitors with appropriate information on transportation lines and their schedules, using appropriate screens, as well as providing local, cultural and sports programs, commercial advertising, etc. (Fig. 11).

Last but not least, a service we would like to mention is the *Pedestrian Drive*, the aim of which is to provide fresh food distribution based on box-lockers [16]. The idea is to convey fresh food to the beach instead of asking clients to go to a supermarket. The pedestrian drive is a variant of the concept of the supermarket drive, specially designed for pedestrians. Located as near as possible to client locations (beaches, sports arenas, etc.) and accessible 24 h a day, it allows ordered goods to be collected at any time. It is able to store not only ambient temperature

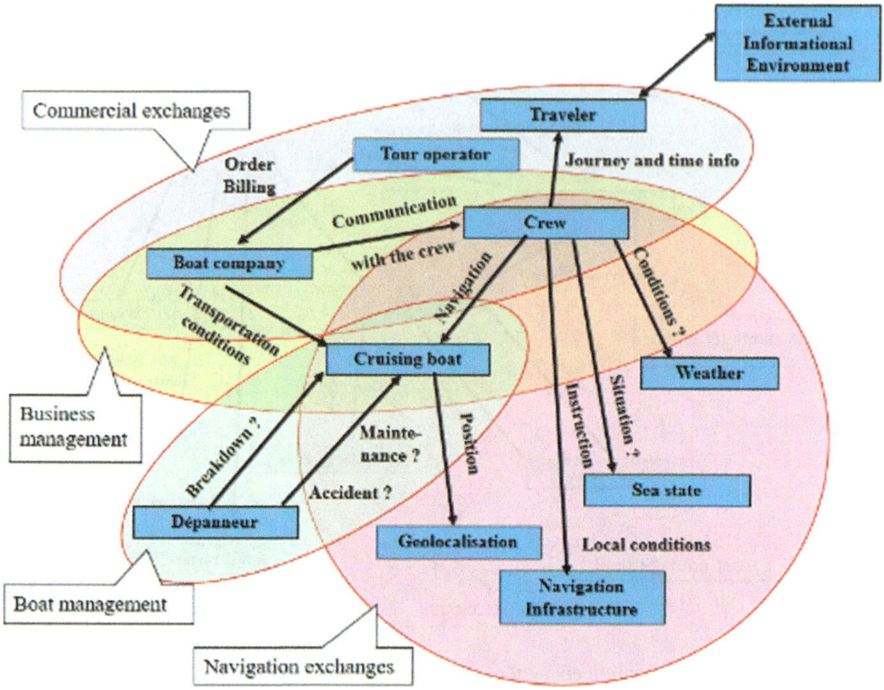

Fig. 7 Cruising boat information ecosystem

products but also fresh and frozen goods. This fresh product box-locker can be either owned by a supermarket firm and totally integrated into the ordering, management, and logistics process of the firm or can be managed independently. This case offers an interesting support system for the shared economy, as the role of the pedestrian drive manager is to manage such use for multiple providers. His/her role is to ensure Internet access for all provider offers and to organize the global supply chain with consumer information on availability of ordered goods in the box-locker. This makes it possible to support the circular economy as a short circuit of agricultural goods (Fig. 12).

For all these identified activities, it is fundamental to define how to collect the data on which they are based and to determine their management and evolution. The following Sect. 4 is devoted to this issue.

Of course, this enumeration is not complete and has been voluntarily limited to mobility aspects.

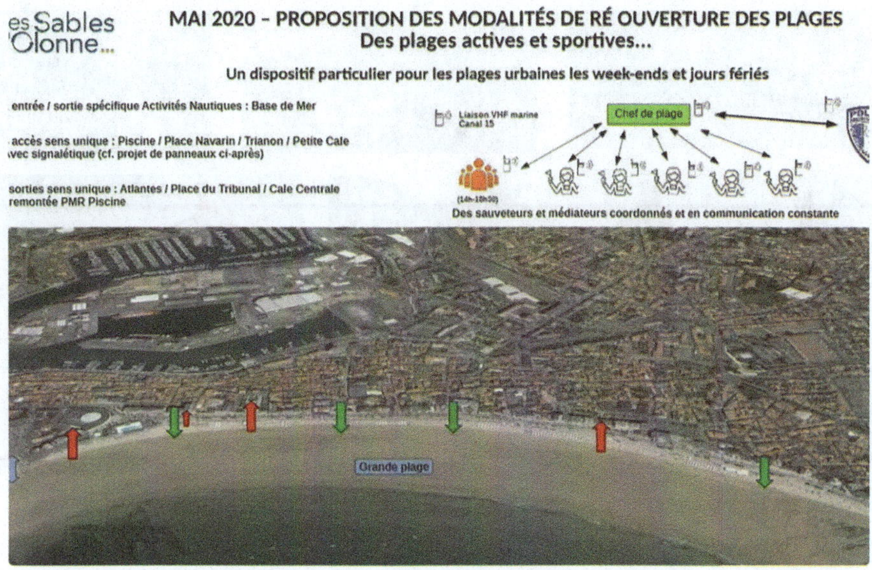

Fig. 8 COVID on the beach [15] © Ville des Sables d'Olonne

Fig. 9 COVID location management on the beach

4 Conceptual View

4.1 *Assistive System Role and Functions*

To provide appropriate services organized in an assistive system, it is important to identify the main aspects of this system. Globally, ITS (Information & Communication Systems) are responsible for considering data collection, storage and management, appropriate manipulations in different working situations, providing communication over short and large distances by wireless and wired networks, in mobility, and interacting with the environment in which active and passive things are organized in IoT (Internet of Things). Two important aspects are HCI (Human-Computer Interaction), and new services based on AI (Artificial Intelligence). The

Fig. 10 Beach and sea zone management, supervision and rescue organization

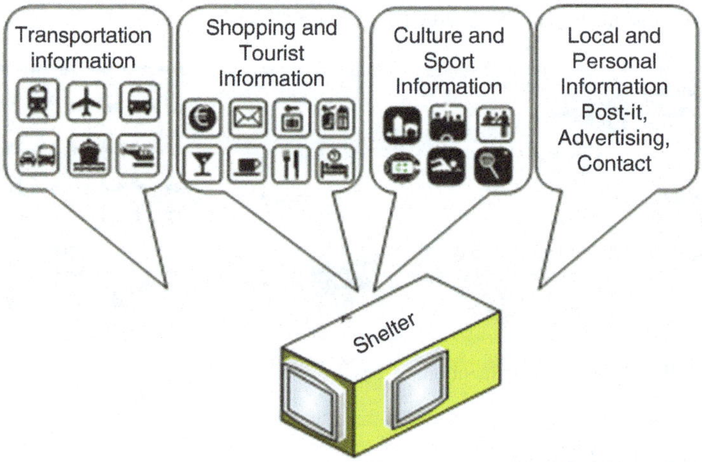

Fig. 11 Bus or boat shelter for inhabitants' information

main HCI objectives are related to multimodal and multichannel communication with various devices and interaction styles such as touch-based and speech-based interactions (Chatbots), as well as AR (Augmented Reality) approaches and soon Metaverse. These interactions can occur either on a desktop computer or with mobile devices (smartphones, connected watches, tablets), allowing individual and collaborative situations, and large, in-the-field, screens or in infrastructure integrated HCI devices, as in the vehicle dashboard.

Fig. 12 Box/lockers for fresh and frozen food [16]

From the AI point of view, Big Data, Open Data, and Machine and Deep Learning are mainly used to procure appropriate information from the huge amount of data available and to synthetize observed behaviors issued from typical data. An interesting summarization is proposed in Fig. 13 [17].

4.2 Mobility Assistive System Platform

To organize all these aspects, it is appropriate to create an intermediation platform, the aim of which is to connect people, services, and even things in ways that have been unthinkable until now. Search engines provide relevant references for people searching for information. Social networks connect users in their environment.

Users' activities on the platform collect the data coming from the field either from specific sensors or from Open Data, and generate secondary data. The latter mainly consist of traces that the platform is able to use and transform for specifically created secondary services.

In order to organize and manage these aspects, we need a system that groups the different services and their applications, the data on which they work, and the relationship, if needed, between the field and the system. We call this system an Assistive System, the aim of which is to manage contextually all smart city applications. Its architecture is organized in four layers (Fig. 14) where:

- Layer 1 is devoted to applications
- Layer 2 is geared toward management of common data and proposing general services
- Layer 3 is working on what is called "Data Vitalization." The idea is to work on data and elaborate new constructions on them that can be used in new applications. An example of this need is to be able to collect and associate data in order to manage situations that were not predictable, such as several cases of rescue for not necessarily easily predictable situations.
- Layer 4 aims at acquiring and distributing the data needed for appropriate working of all applications concerning the coastal smart city.

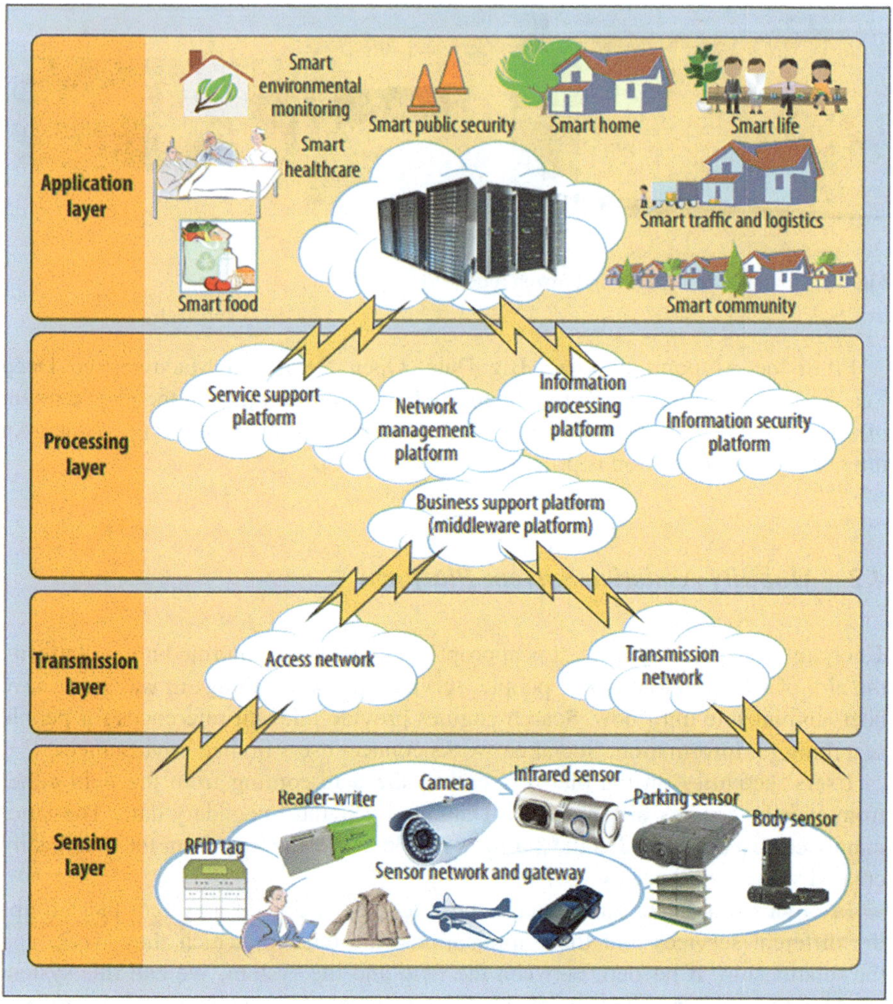

Fig. 13 Smart city architecture summarized by Liu and Peng [17]

- In order to take into account the evolution of this assistive system, an appropriate user interface, called the Supportive User Interface, is provided, allowing maintenance staff to facilitate the modifications requested easily. The "mashup" technique can be used for this purpose [18].

A generic intermediation platform needs to be adapted to different working situations. The objective is to provide appropriate collection of data, services, and acquisition sensors and valuators, which are able to work together in a vitalization approach and create adapted User Interfaces allowing appropriate interactions. The Mashup approach is an interesting way to support these adaptations. Mashups

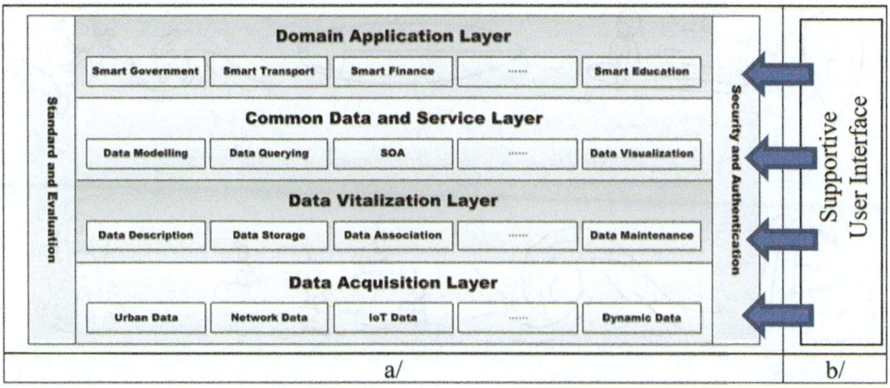

Fig. 14 (a) Architecture of an Intermediation platform with Data Vitalization [19], (b) Supportive User Interface [20]

are defined as "the software engineering approach, which is able to construct by assembling and combination of several existing functions new applications" [21].

The Mashup architecture is made up of three elements according to Merrill cited by [21]: Data, Services, and User Interface. Mashup aims at the composition of a three-tier application: (1) Data (data integration), (2) Application logic (process integration), and (3) User Interface (presentation integration). Integration of heterogeneous data sources uses two main technologies: web services and Mashup.

Application mashuping in the context of intermediation platforms is a much-appreciated approach for creating appropriate applications based on reuse of existing ones. It is based on a four-layer architecture (14 a). To update or adapt data and services, a particular User Interface is used called "Supportive UI," the aim of which is to introduce appropriate adaptations concerning data, services, IoT exchanges, and HCI. This supportive User Interface, which is also Mashup application creation- or evolution-oriented, proposes a programming interface for experienced developers, and a visual programming-oriented approach for experienced users (14 b).

4.3 Data Identification and Dynamicity Management

For all identified activities and applications, it is important to identify the data needed and to determine their evolution, i.e., dynamicity. We identified several levels and providers of data dynamicity.

The first, basic level, is *totally static*, which means not smart, but useful for a large majority of city users (on the whole or by category): physical panels, boards and signs, as well as circulation signs. A contextual example can be city maps, circulation signs, and so on (Fig. 15).

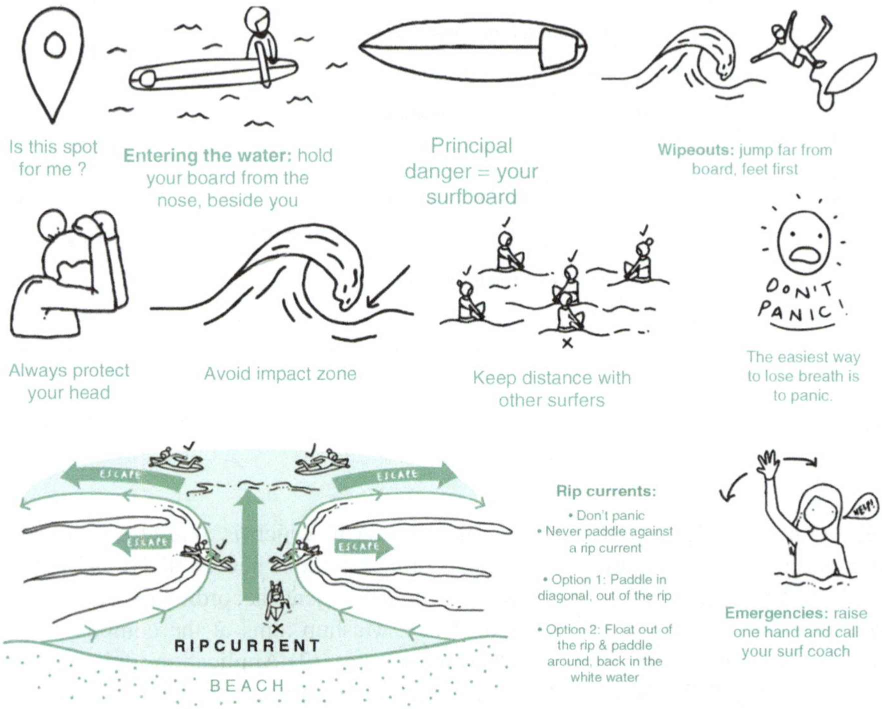

Fig. 15 Surf safety physical panel [22]

The second level, with very *low dynamicity*, is based on boards with QR codes. In this case, in the field (city), the information indicator is static, and dynamicity is provided by the information sent to the user's smartphone from the website address encoded in the QR code. This approach is appropriate for optional information, which is accessed only by interested users and known only to them, in the form of historical information for example, but also for commercial and advertising purposes.

The third level of dynamicity is characterized by the need to *propagate information that is of general interest and needs to be commonly known*. The solution is to propagate it directly on users' smartphones or on GPS vehicle screens. Another solution is to use public active screens and information boards. This is the case for circulation information for dynamically managed circulation lanes such as we described earlier. This kind of dynamicity is usually determined by external decision centers working on collected or predefined data.

The fourth level of dynamicity is *contextual dynamicity*, i.e., it is related to particular cases or situations that can be produced by an action on a local device, manually or automatically. A classic example is a sender located on a bus, which activates the opening of a barrier accessing a restricted portion of the road. This kind of sender (or remote control) can be either integrated on the vehicle (bike, car, bus, boat, etc.) or used manually by the user, often in emergency situations. These devices are attributed by the relevant authority. The actuator associated with the remote control can be either mono-located and mono-answer acting on a single door, barrier, etc., or integrated in a system assisting more sophisticated behaviors. Management of this dynamicity requires appropriate association between sender(s) and receiver(s) in order to organize the corresponding behaviors.

The highest-level concerns *emergency and rescue situation dynamicity*, mainly related to weather and environmental situations, such as floods, landslides, tsunamis, ecological and industrial problems, and fires. In these situations, the objective is to collect appropriate information and transmit it to the relevant organizations in charge of determining appropriate actions and informing those concerned (both those having to take action and those impacted).

From the data collection point of view, it is important to dispose of appropriately distributed sensors in water, air, and elsewhere for fire detection, which are able to transmit the appropriate data to the rescue center. The synthesis of these data is then elaborated, and an emergency signal is propagated to all information media, both individual and collective, in the city with vocal and/or visual announcements indicating the actions to take (leave the sea, put on a mask, move to an appropriate location, etc.).

4.4 Data Collection and Elaboration

As analyzed in the previous section, data availability is the main problem in the provision of appropriate solutions. For the first levels of dynamicity, work can be elaborated progressively and be used relatively long term. For greater levels of dynamicity, three different approaches can be used.

The first is totally related to the local context. The aim of local senders addressing local receivers is to obtain local and appropriate answers, such as opening of a read section that is managed.

A more general solution is to be able to share available data that are collected by communicating devices, namely, senders and receivers, on site in the field, the city, and its neighborhood and open to external users for their applications and services. This approach, which is called Open Data, presented below, constitutes major progress in general utilization of available data.

Another interesting approach is based on evolution of collected data by the acquisition of new, more recent data and their integration in the elaborated behavior model. This approach is based on a very active research field called Machine and Deep Learning, which objective is to aggregate collected data and create progressively behavioral model synthesizing observed situations.

4.5 Open Data

One of the possible sources of information is the access to open data published by different operators, such as municipalities, road operators, etc. These data can be either historical data synthetized and giving general tendencies or real-time data, indicating what is happening now or a few minutes ago.

Open Data [23] is the idea that some data should be freely available to everyone to use and republish as they wish, without restrictions from copyright, patents, or other mechanisms of control [24]. The goals of the open-source data movement are similar to those of other "open(-source)" movements such as open-source software, hardware, open content, open specifications, open education, open educational resources, open government, open knowledge, open access, open science, and the open web. Paradoxically, the growth of the open data movement is paralleled by a rise in intellectual property rights [25]. INSPIRE (Infrastructure for Spatial Information in the European Community) is an EU initiative to establish an infrastructure for spatial information in Europe, which is geared to help make spatial or geographical information more accessible and interoperable for a wide range of purposes supporting sustainable development.

4.6 Machine Learning and Deep Learning

Deep Learning is hugely popular today. The past few decades have witnessed its tremendous success in many applications. Academia and industry alike have competed to apply deep learning to a wider range of applications due to its capability to solve many complex tasks while providing state-of-the-art results [26]. In some situations, Deep Learning can be used to analyze existing data, leading to interesting solutions. We can take intelligent navigation and travel planning as examples. For intelligent navigation, the intermediation platform considers the starting position and the destination position, combines the time information, meteorological information, and other interfering factors before providing the most appropriate one or more recommended paths. As for travel planning, the focus is to intelligently analyze the user's intention of their behavior and finally make travel recommendations based on the tags they are interested in and their browsing history.

In practice, we can use the RNN structure in path planning, including intelligent navigation, travel for the disabled, Parcel post, and so on, while the CNN model can be used for vehicle recognition to assist smart parking. For some scenarios with recommendation tasks, such as goods transportation and travel planning, we can use multi-layer perceptron, LSTM, and other models [27].

4.7 Data Model Structure and Access Management

All assistive services need to use, produce or evolve data, which can be personal, ephemeral, long life, public, ... They are used either in elementary services or more sophisticated multiservices. If ephemeral data are specifically collected, elaborated, and used each time, the data with long-term duration must be stored in order to facilitate their use. They can be stored individually, or by theme, in what is called data silos. A more general integration could also be possible, but for smartification, it seems important to still be open to receive new applications using existing or new data. In this way partial integration is an appropriate solution with thematic silos and appropriate API (Application Programming Interfaces) managing data accesses. Inter-silo relations are in the same situation with appropriate major exchanges and the necessity to improve less frequently used operations. Total integration seems impossible to obtain, as the global system is not closed but rather open-ended, able to receive (each day) new services and applications supporting new visions of digitalized/virtual tourist activities.

Introduction of a new digitalized service can be either independent or need to collaborate with other services already present. These services must provide an API (Application Programming Interface), a standardized approach for their interconnection.

5 Design Methodology

5.1 Design Approaches

As we explained in the previous sections, the process consisting in transforming a city into a smart city usually starts by identifying the actors concerned, as well as the necessary services and associated data.

To design a Smart City, it is possible to proceed in several ways [25]. The easiest is to create from scratch an application and deploy it. When several applications are proposed, a more system-oriented approach is suggested, which either can be proposed or must be elaborated (Fig. 16).

Several design methods can be used, depending on the context in which they evolve. In the case of Smart City system design from scratch, the *top- down approach (1)* seems appropriate. In this case, the study starts with identification of potential users and corresponding services by a discussion with them in order to identify their wishes and expectations. Naturally, the next steps are identification of data on which these selected services can work and, finally, the sensors and actuators needed to obtain an appropriate relationship with the field. These different aspects are implemented progressively and place in the Assistive System Platform described earlier.

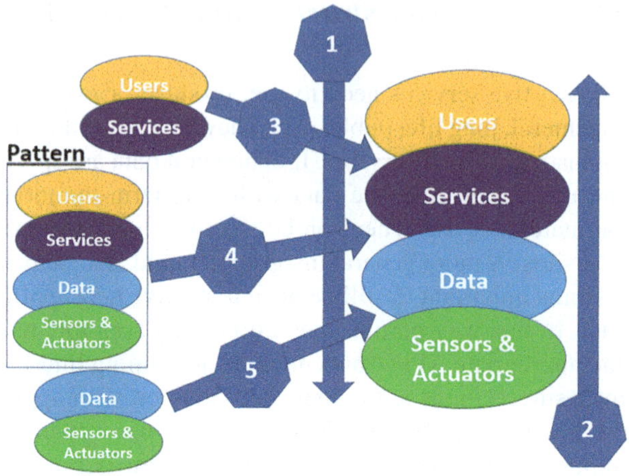

Fig. 16 Design and deployment approaches

Conversely, the *bottom-up approach (2)*, could be appropriate if the objective or main question is which services can be provided to potential users on the bases of available data. In this case, the working process starts with identification of in-the-field devices and corresponding data provided by them. Also, services and potential users are determined later. This environment-based approach is appropriate if the starting point is composed of a large scope of problems, and if the set of potential users (and services) is open-ended.

Naturally, between these "extreme" approaches an *extension-oriented approach* is also possible, and may probably be most commonly used. Its characteristic is its adaptation to extension-oriented situations, i.e., to consider *new users, new services, and/or new data integration to an existing Smart City Assistive System (3)*.

In this case, two sub-approaches can be distinguished, in the same way as for the "from scratch" approach: top-down or bottom-up with use of different starting points: either from users and services or from data, sensors, and actuators needed. In these two cases, the objective of this integration process is to identify the new set of actors and homogenize services, data, sensors, and actuators to avoid redundancy and incoherence that are difficult to manage during platform life.

Another approach is based on integration and reuse of *on-the-shelf applications*, constituting *patterns (4)*, which have been created earlier and proposed as pattern solutions. In this case, the integration process could also contain homogenization considering all the pattern characteristics: User + Services + Data + Sensors & Actuators to reach a homogeneous system.

The last but not the least approach is based on *Open Data as starting point (5)*, and integration or elaboration of appropriate services using them.

5.2 *Case Studies: Smartification of a Coastal City*

5.2.1 Top-Down Approach: City Smartification

An example of the Top-down approach in the Coastal City is the objective to work with all concerned persons and determine main orientations of "smartification" from their viewpoints in the context of mobility as we summarized in the introduction. We can add to mentioned services the objective of creating park-and-ride systems outside the city with automatic public transportation to the town and the beaches, to optimize traffic organization on the ground, in the harbor and the sea, as well as management of beach activities. In this way it seems appropriate to begin by the discussion with potential users and collect expected services. Identification of needed data, their collection, with appropriate devices, and long-time management constitute the new step as well as the way to propagate these data to the users by appropriate supports as wireless networks.

From application point of view, we consider it as a Top-down designed one, starting by general discussion and progressive refinement of all needed aspects (treatments, data, user interfaces, in-city devices, . . .).

5.2.2 Bottom-Up Approach: Emergency and Rescue Services

Concerning the Bottom-up approach, an interesting example is to provide an efficient warning system based on in-the-field devices (sea, water, air, lighthouse, stormy and howling wind, etc.), propagating emergency and rescue information at all levels (global, individual, visual, sound, digital, etc.).

From the application point of view, we consider it as Bottom-up design approach, starting by the capacity of the environment to provide appropriate information. Naturally, if this step is not satisfactory, it will be necessary to continue with a more or less Top-down approach.

5.2.3 Bottom-Up Approach: Traffic Management

Organization of traffic is a very important aspect. Of course, in the low season, we can consider that static allocation of the circulation map is appropriate. However, in the high season, dynamicity aims at ensuring better traffic speed by introducing several restrictions during different periods of the day with authorized or prohibited vehicles. Circulation restrictions can vary frequently between buses, trucks, tourist vehicles, and so on. It is also important to take into account the main objective of each zone in the city. In tourist zones, such a variation is natural. In industrial zones, mainly dedicated to boat construction and maintenance, priority must be given to the fluidity of this activity. Such dynamicity is either based on collected and

Fig. 17 The GPS wristband saving lives [22]

preprocessed data (behavioral patterns) or on real-time data. This is then propagated to the assistive system and drivers in an appropriate manner, either by electronically driven signs and/or by propagation of corresponding information to the in-vehicle screens, smartphones, and GPS applications.

From application point of view, we consider it as bottom-up designed one, based mainly on available data, which can be received and observed by the users.

5.2.4 Pattern Use: Data for Beach Mobility

As we explained earlier, concerning management of mobility on beaches and on the sea in recreation areas, static information is available, mainly to indicate authorized and banned zones and expected users' behaviors. In this case, security observations and rescue actions are based on visual observations.

In a more dynamic and secure approach, swimmers and kite surfers are equipped with appropriate transceivers indicating their geographical positions (Fig. 17). These devices are able to activate emergency demands. Life guards are fitted with a receiver designed to observe the movements of everyone in the beach zones and beyond in order to observe users' behaviors and contact each one in order to clarify their situation. They can indicate inappropriate behaviors, such as use of an inappropriate zone or ask the swimmer for the reason for his/her emergency signal and trigger appropriate action. From the application point of view, we consider it as a specific one, which is mainly based on a pattern structure:

- Specific user-interface on specific device or generic mobile device (smartphone, digital watch or tablet)
- Specific application calculating user's position and receiving emergency information
- Specific data and GPS information
- Using ad hoc or standard wireless network access

5.2.5 Pattern Approach Based on Open Data: Information Shelter

As an example of the service based on open data, we choose a shelter. We initially proposed a bus shelter [28], but its adaptation to other situations, such as beach shelter, harbor shelter, public market shelter is not difficult.

The shelter can be located downtown, near the carpark, on the harbor, and near the beaches. Its objective is to provide potential passengers and visitors with appropriate information on transportation lines and their schedules, using appropriate screens, as well as providing local, cultural and sports programs, commercial advertising, etc., together with, if needed, physical services (shower or changing cabin). The information part is mainly based on presentation of generally available information issued from open data.

5.2.6 Pattern-Based Approach: Pedestrian Drive

As an example of a pattern-based approach, we can present what we call the Pedestrian Drive, the aim of which is to provide fresh food distribution based on box-lockers [16]. The idea is to convey fresh food on the beach instead of asking clients to go to a supermarket. The Pedestrian Drive is a variant of the concept of the supermarket drive, specially designed for pedestrians. Located as near as possible to the client locations (beaches, sports arenas, etc.) and accessible 24 h a day, it allows ordered goods to be collected at any time. Equally, it is able to store not only ambient temperature products but also fresh and frozen goods. This fresh product box-locker can be either owned by a supermarket firm and totally integrated into the ordering, management, and logistic process of the firm, or can be managed independently. This case is an interesting support system for the shared economy, as the role of the pedestrian drive manager is to manage such use for multiple providers. His/her role is to ensure Internet access for all provider offers and to organize the global supply chain with consumer information on availability of ordered goods in the box-locker. In this way, the circular economy as a short circuit of agricultural goods can be supported. Its characteristic, in which we are interested, is its independence from the implementation location. That is why we present it as a pattern of autonomous equipment completely characterized by what we call Pattern: User + Services + Data + Sensors & Actuators, which are easy to position at the chosen location.

5.2.7 Add Extension: Boat Management

Smartification of boat management is concerned by the circulation which can be either limited to the local area, or generalized to the overall management and supervision of boats in the harbor and on the sea. The ecosystems for fishing and cruising boats has been presented earlier, schematized in Figs. 6 and 7, main actors, services and data. The principle is to identify all concerned actors and situations and indicate their behaviors, their interrelations, and contributions to appropriate trip and work conditions. The design approach is composed of "Patterns" for well identified behaviors and needs and "Add extension" for more new behaviors.

5.2.8 Add Extension: Beach and Sea Mobilities

A very important "smartification" is related to beach and sea mobilities. This can be either the main aspect of the overall study or added separately. On the beach several activities can take place, more or less active in nature: from sunbathing, walking along the seashore, practicing different sports or family activities. In all cases, it is important to take into account other beach users, in order to avoid accidents and disturbing them. During the recent COVID period, for many French beaches, city mayors at local level or, more globally, government authorities decided on several levels of rules: prohibit access to the beaches, authorize only "dynamic use of the beach" (no sitting or sunbathing, only walking and running). This also included respect of distances between groups of users, either natural or physically materialized zones with the possibility of reserving them.

On the sea, the development of activity zones is the usual approach leading to creation of specific zones with supervised swimming, diving, snorkeling activities, as well as sailing, yachting, surfing, kitesurfing, with either visual supervision or more instrumented electronic monitoring. In this case, each actor (swimmer, sailor, kite surfer) has at their disposal a connected watch or more sophisticated device able to communicate their position and related situation (OK, emergency call, etc.). On the beach, lifeguards are equipped with a receiver capturing the observed situation and able to trigger rescue actions.

5.2.9 Machine and Deep Learning Use

We would also briefly present a possible use of Machine and Deep Learning technics. Our case study is sea behavior oriented. Its objective is to study evolution in sea behavior over the season and, in particular, weather conditions as the wing and tide coefficient changes to inform swimmers and/or kiters of the organization of the supervised zone. By accumulating the data of different periods, it would seem possible to indicate foreseeable sea behavior (Fig. 18) [29].

5.2.10 Smart Services for Disables

In the evolution of city services, it seems important to not forget services for disabled people. As examples we can mention dimension three services on which we previously worked [31]:

- Indoor navigation (technology based) is a service that allows disabled persons to move inside of unknown building by the aid of active or passive wireless

Fig. 18 Surf safety panel issued from history observation [30]

network creating appropriate exchanges between close environment and current real position of moving person.

- Outdoor navigation (technology based) is an outdoor version of the same service, allowing moving persons to navigate in real life 3D environment, based on wireless network and in the environment located devices (senders and receivers) participating to Internet of Things (IoT).
- Escort disabled (human accompanying) is another approach in which the mobility of disabled person is provided by the cooperation between him/her and another person moving him/her in the environment. Different approaches are possible, either totally human based or supported by technology: appropriate user interface enabling to organize city displacement in time and location and also physical meeting points in which the escorting person and the escorted one can meet, in order to carry out the move using public transportation.

6 Conclusion

In this chapter we studied how a coastal city can evolve to become a Smart Coastal City. The process is mainly based on the study of the actors concerned (inhabitants, tourists, workers, government authorities, etc.) and the corresponding services for each of them and globally. The specificity of a coastal city with activities related to the sea, industry, and tourism were the main aspects for this study. Our reasoning was based on the existence of two seasons (the high season – tourism oriented – and the low season – inhabitant oriented) in which different services are needed and can be used differently.

We proposed what we consider to be major transformations, either specific to the coastal city or adapted from the smart city in general. We identified expected services, associated data, corresponding user interfaces, individual or more common, and we described the links (passive or active) with the field (part of the city concerned) and the need for in-the-field indicators, senders, and receivers. The dynamicity of the functional periods was also studied. We only briefly presented technological aspects (assistive system as an intermediation platform, open data use, and Machine and Deep Learning techniques), which we consider out of scope of this conference.

We presented our methodology allowing to design different services or applications proposing five approaches and we presented several case studies using these approaches.

We also shortly mentioned our interest for assistance to disabled people. In [31] we presented several solutions applicable in Smart Cities and that can be transposed easily to this context. Of course, new solutions, specific to the coastal context (beaches, sea, harbor, boats) can be proposed. This is our objective in the near future, together with the discovery of other Coastal Smart City services and applications. Naturally, our study is not complete. Its objective was to propose a design methodology proposing different approaches for "smartification" of cities in a particular context of the Coastal City. We voluntarily limited our study to mobility aspects. and avoided two very important problems, namely, adaptation of cities during "smartification" to the elderly and, in particular, to disabled people.

Acknowledgments This work is a collaboration between one French team located in Lyon and one Chinese team located in Beijing. We thank our partners and their students and researchers for their contributions. This study is partially supported as regards China by the National Key R&D Program of China (No. 2019YFB2102200) and the National Natural Science Foundation of China (No. 61977003). As regards France, the study was mainly supported by the French Ministry of Ecology (No. 09MTCV37 and PREDIT-G02 CHORUS 2100527197).

References

1. Caragliu, A., Del Bo, C., Nijkamp, P.: Smart cities in Europe. J. Urban Technol. **18**(2), 65–82 (2011)
2. Yin, C., Xiong, Z., Chen, H., Wang, J., Cooper, D., David, B.: A literature survey on smart cities. Sci. China Inf. Sci. **58**, 100102:1–100102:18 (2015). https://doi.org/10.1007/s11432-015-5397-4. Science China Press and Springer-Verlag Berlin Heidelberg 2015 info.scichina.comlink.springer.com
3. Kolski, C.: Human-Computer Interactions in Transport. Wiley (2011)
4. Hosaka, T.A.: Japan creating 'smart city' of the future. San Francisco Chronicle. Associated Press, 11 Oct (2010)
5. Ng, P.T.: Embracing emerging technologies: the case of the Singapore Intelligent Nation 2015 vision. In: de Pablos, P.O., et al. (eds.) Regional Innovation Systems and Sustainable Development: Emerging Technologies, pp. 115–123. IGI Global (2011). https://doi.org/10.4018/978-1-61692-846-9.ch008
6. Giffinger, R., Gudrun, H.: Smart cities ranking: an effective instrument for the positioning of the cities? Arch. City Environ. **4**(12), 7–25 (2010)
7. Washburn, D., Sindhu, U.: Helping CIOs Understand "Smart City" Initiatives. Forrester Research, Inc, Cambridge (2010)
8. Nam, T., Pardo, T.A.: Conceptualizing smart city with dimensions of technology, people, and institutions. In: Proceedings of 12th Annual International Digital Government Research, College Park, vol. 12–15, pp. 282–291 (2011)

9. Isla, F.I.: From touristic villages to coastal cities: the costs of the big step in Buenos Aires. Ocean Coast. Manag. **77**, Elsevier (2013). https://doi.org/10.1016/j.ocecoaman.2012.02.005
10. Chalon R., Yin, C., David, B.: Assistive systems for special mobility needs in the Coastal Smart City. In: HCI International 2022 24th International Conference on Human-Computer Interaction, 1 juillet 2022, Full Virtual Conference (États-Unis), pp. 489–506. https://doi.org/10.1007/978-3-031-04987-3_33. HAL : hal-03658281. (2022)
11. Chuai, X., Huang, X., Wu, C., Li, J., Lu, Q., Qi, X., Zhang, M., Zuo, T., Lu, J.: Land use and ecosystems services value changes and ecological land management in coastal Jiangsu, China. Habitat Int. **57** (2016). https://doi.org/10.1016/j.habitatint.2016.07.004
12. Wang, C., David, B., Chalon, R., Yin, C.: Dynamic road lane management study: a Smart City application. J. Elsevier Transp. Res. E: Logist. Transp. Rev. **89**, 272–287 (2015). https://doi.org/10.1016/j.tre.2015.06.003
13. Patier, D., David, B., Deslandres, V., Chalon, R.: A new concept for urban logistics: delivery area Booking. In: The Eighth International Conference on City Logistics, Eiichi Taniguchi, Russell G. Thompson ed. Bali, Indonesia. pp. 99–110. Procedia Social and Behavioral Sciences 125. Elsevier. ISSN 1877-0428 (2014)
14. David, B., Xu, T., Jin, H., Zhou, Y., Chalon, R., Zhang, B., Yin, C., Wang, C.: User-oriented System for Smart City approaches. In: The 12th IFAC/IFIP/IFORS/IEA Symposium on Analysis, Design, and Evaluation of Human-Machine Systems, August 11–15, 2013, Las Vegas, Nevada, USA. pp. 333–340. IFAC/Elsevier. ISSN 1474-6670 (2013)
15. https://france3-regions.francetvinfo.fr/pays-de-la-loire/coronavirus-plages-baule-pornichet-sables-olonne-nouveau-accessibles-1828366.html
16. David B., Chalon R.: Box/Lockers' contribution to collaborative economy in the Smart City. In: 2018 IEEE 22nd International Conference on Computer Supported Cooperative Work in Design (CSCWD 2018), 11 May 2018, Nanjing (Chine), pp. 802–807 (2018), https://doi.org/10.1109/CSCWD.2018.8465151
17. Liu, P., Peng, Z.: China's smart city pilots: a progress report. Computer. **10**, 72–81 (2014)
18. Grammel, L., Storey, M.: An End User Perspective on Mashup Makers. University of Victoria. Technical Report DCS-324-IR (2008)
19. Xiong, Z.: Smart city and data vitalisation. In: The 5th Beihang Centrale Workshop, 7 Jan 2012
20. David, B., Chalon, R.: Orchestration modeling of interactive systems. In: Jacko, J.A. (ed.) HCI 2009. LNCS, vol. 5610, pp. 796–805. Springer, Heidelberg (2009). https://doi.org/10.1007/978-3-642-02574-7_89
21. Atrouche, A., Idoughi, D., David, B.: A mashup-based application for the smart city problematic. In: Kurosu, M. (ed.) HCI 2015. LNCS, vol. 9170, pp. 683–694. Springer, Cham (2015). https://doi.org/10.1007/978-3-319-20916-6_63
22. https://barefootsurftravel.com/fr/livemore-magazine/aspects-de-securite-en-surf
23. Auer, S.R., Bizer, C., Kobilarov, G., Lehmann, J., Cyganiak, R., Ives, Z.: DBpedia: a nucleus for a web of open data. In: The Semantic Web Lecture Notes in Computer Science, vol. 4825, p. 722 (2007). https://doi.org/10.1007/978-3-540-76298-0_52. ISBN 978-3-540-76297-3
24. Kitchin, R.: The Data Revolution, p. 49. Sage, London (2014) ISBN 978-1-4462-8748-4
25. https://en.wikipedia.org/wiki/Open_data
26. Paul, C., Jay, A., Emre, S.: Deep neural networks for YouTube recommendations. In: RecSys'16, pp. 191–198 (2016). https://doi.org/10.1145/2959100.2959190
27. Huang, P.-S., He, X., J. Gao, Deng, L., A. Acero, L. Heck: Learning deep structured semantic models for web search using click through data". In: CIKM 2013: Proceedings of the 22nd ACM International Conference on Information & Knowledge Management, pp. 2333–2338 (2013)
28. David, B., Chalon R.: Hotspot based mobile web communication and cooperation: ABRI+ Bus Shelter as a hotspot for mobile contextual transportation and social collaboration, SWWS'2010 (Semantic Web and Web Services) as part of WorldComp'2010 Conference, Las Vegas (2010)

29. David, B., Chalon, R., Yin, C.: Design methodology for "Smartification" of Cities: principles and case study. In: IEEE 25th International Conference on Computer Supported Cooperative Work in Design, 6 May 2022, Hangzhou (Chine), pp. 1486–1491. https://doi.org/10.1109/CSCWD54268.2022.9776257. HAL : hal-03658035. (2022)
30. https://www.dial.help/
31. Yin, C., David, B., Chalon, R., Sheng, H.: Assistive systems for special needs in mobility in the Smart City. In: HCI International 2020 22nd International Conference on Human-Computer Interaction (2020), https://doi.org/10.1007/978-3-030-50537-0_27

Urban Climate and Heat Mitigation in Coastal Cities

Mattheos Santamouris and Konstantina Vasilakopoulou

1 Introduction

Coastal cities face important environmental and energy problems caused by regional and global climate change. Overheating is the major problem encountered in many coastal cities combined with other serious environmental and energy impacts caused by the higher urban temperatures. Regional climatic change affects the energy consumption of buildings for cooling, increases the peak electricity demand considerably and creates the need for additional power plants that operate just for a few days per year. Moreover, regional climatic change rises the cost of electricity, reduces the efficiency of nuclear and thermal power plants, increases the concentration of harmful pollutants and especially of the toxic ground level ozone, causes important health problems and exacerbates heat-related mortality and morbidity, while affecting the survivability of low-income households [1].

Compared to mainland cities, coastal cities benefit considerably from the sea breeze developed mainly during the early afternoon hours because of the important temperature difference between the land and the sea. Sea breeze brings cool fresh air to the urban environment, making it the most important cooling mechanism for coastal cities. The length of the penetration of the sea breeze in a city depends on many parameters like the magnitude of the temperature difference between the land and the sea and the local landscape [2].

To improve the climatic conditions and enhance the cooling potential of sea breeze in coastal cities, important mitigation technologies can be implemented

M. Santamouris (✉)
Anita Lawrence Chair High Performance Architecture, School of the Built Environment, Faculty of Arts, Design and Architecture, University of New South Wales, Sydney, NSW, Australia
e-mail: m.santamouris@unsw.edu.au

K. Vasilakopoulou
School of the Built Environment, Faculty of Arts, Design and Architecture. University of New South Wales, Sydney, NSW, Australia

S. Th. Rassia (ed.), *The Blue Book*, https://doi.org/10.1007/978-3-031-48831-3_6

at the city scale. Mitigation technologies involve the use of innovative reflective and fluorescent materials that present sub-ambient temperatures under the sun, appropriate urban greenery, evaporative systems, solar control devices and use of cooling systems linked to low temperature environmental sinks [3]. Analysis of the cooling performance of a large number of large-scale heat mitigation projects around the world has shown that it is possible to decrease the peak temperature of cities up to 2.5 °C, while the implementation of the super cool materials is expected to further increase the cooling potential up to 4 °C [4, 5].

Implementation of mitigation technologies in cities offer important benefits. Heat mitigation can reduce the electricity demand for cooling up to 50% [6], decrease the magnitude of heat-related mortality and morbidity up to 40% [7], reduce the concentration of harmful pollutants substantially [8] and improve the survivability levels of low-income population in cities [9].

2 Climate of Coastal Cities: Interaction of Sea Breeze with Urban Heat Island Characteristics

Previous studies have documented the interaction of sea breeze in coastal cities with the urban heat island characteristics [10–12]. It is a common conclusion that in coastal cities the presence of sea breeze during the daytime of warm periods is decreasing the magnitude of the urban overheating mainly because of the transfer of cool air into the city [13].

Several experimental and numerical studies have analysed the characteristics of the interaction between urban heat island and sea breeze in coastal cities. Most of the studies agree that urban heat island affects the characteristics of the sea breeze, and that the urban heat island is well developed and persists even with the presence of the sea breeze [14]. As shown in [11], urban heat island in Tokyo, Japan, creates a stagnation region over the city that can delay the flow of sea breeze into the city. This results in less deep penetration of the sea breeze in the inland area and spatially limits its cooling potential in the coastal area considerably [15]. Studies in Sao Paolo, Brazil and elsewhere, report that the urban heat island increases the propagation speed of the local sea breeze towards the centre of the city [10, 16], but due to the serious convergence zone created over the city, the front of the sea breeze may stall over the city for a couple of hours and start to propagate when the magnitude of the urban heat island is very small [10]. Such a phenomenon is more prominent in larger cities located in a substantial distance from the coast where the urban heat island phenomenon has time to develop [17]. Slightly different results are reported in [12], for the city of New York, USA, where experimental and numerical studies concluded that during the warm period, the strength of the sea breeze is reduced and delays the local circulation induced by the urban heat island.

Other studies carried out mainly in Asian cities, confirmed the previously reported conclusions. According to studies performed in Korea, sea breeze seems to have an important impact during the spring and summer period but it is very weak or it disappears during the rest of year [18, 19]. Similar results are reported from experiments performed in Sendai and Tokyo, Japan, India, Jeddah, Tel Aviv and Pakistan [20, 21].

The possible impact of tall buildings on the characteristics of sea breeze in coastal cities is investigated by many studies. It is found that dense buildings and the roughness created by the buildings have a useful impact and contribute to pull sea breeze down to the ground level of the city [18].

3 Heat Mitigation and Adaptation Techniques in Coastal Cities

To counterbalance the impact of high urban temperatures and improve the comfort and environmental conditions in cities, innovative heat mitigation technologies have been developed and implemented. Heat mitigation technologies aim to decrease the strength of heat sources and improve the cooling performance of heat sinks.

Heat mitigation technologies involve the use of innovative materials, urban greenery, evaporative and solar control devices and systems associated with the use of low temperature environmental sinks [3].

The materials used for the envelope of urban buildings and other city structures determine at large the thermal budget of cities. Materials absorb solar radiation and emit infrared radiation. Dark materials present a much higher absorptance to solar radiation and exhibit a much higher surface temperature than the reflective light colour materials. The development of white highly reflective materials in the whole solar spectrum, as well as of the coloured materials reflecting in the infrared solar spectrum, has boosted mitigation technologies, and has offered a substantial cooling potential for cities [22–24]. Numerous applications of the developed reflective materials all around the world have shown that they can contribute to decrease the peak temperature of cities up to 1.5 °C [25]. The research on a second generation of advanced heat mitigation materials has resulted in the development of thermochromic materials able to change their colour and surface temperature as a function of the outdoor ambient temperature [26]. Thermochromic materials exhibit up to 20 °C lower surface temperature than the conventional materials of the same colour and are appropriate for cities presenting both heating and cooling needs [27]. However, the recent development of photonic daytime radiative cooling materials has offered tremendous new opportunities for heat mitigation [28]. Photonic radiative materials present an extremely high reflectance to solar radiation, of the magnitude of 90–95%, combined with a high emissivity value in the atmospheric window [29]. As a result of their optical characteristics, they can exhibit sub-ambient temperatures under the sun while they may decrease the peak ambient temperature in cities up to 4.5 °C [30].

Increase of the green infrastructure in cities highly contributes to the decrease of the high ambient urban temperatures. Greenery reduces the thermal balance of cities through evapotranspiration and solar control [31]. Increase of the urban green infrastructure can decrease the peak daytime temperature in cities up to 1.5 °C, and the night temperature up to 10 °C [8]. Unfortunately, the evapotranspiration rate of greenery is seriously reduced under high ambient temperatures while the emission of the toxic BVOC by greenery may increase the concentration of harmful pollutants like the ground level ozone [32].

Combined implementation of mitigation technologies resulted in a reduction of the peak temperature in cities up to 2.5 °C. It is reasonably expected that the implementation of the innovative heat mitigation technologies may boost the cooling potential up to 4–5 °C [4].

4 Impact of Heat Mitigation on the Energy Consumption, Health, Environmental Quality and Quality of Life

Implementation of heat mitigation technologies in cities is found to have a serious impact on human life. Decrease of the ambient temperature contributes to considerably decrease the cooling energy consumption of buildings, lower the magnitude of heat related mortality and morbidity, reduce the concentration of harmful pollutants, improve the survivability of low-income households and boost human productivity [33].

Several studies have assessed the impact of the urban albedo rise on the energy consumption of buildings. It is found that the use of reflective materials at the city and building scale in Sydney, Australia contributes to reduce the cooling energy consumption between 13% and 21%, depending on the characteristics of the building and the local conditions [34]. Similar studies carried out for the city of Phoenix, USA, found that a potential increase of the urban albedo by 0.1, 0.25 and 0.4 may decrease the buildings cooling consumption by 5.2%, 8.5% and 13.1%, respectively [35].

Increase of the green infrastructure (GI) in cities is also found to reduce the cooling energy demand of buildings. It is estimated that by increasing the GI by 30% and 60% in Riyadh, Saudi Arabia, the cooling energy demand may decrease by 3.9% and 10.6%, respectively [6].

Higher urban albedo values are found to result in a considerable decrease of the heat-related mortality (HRM) in a city. Analysis of numerous studies has found that the resulting decrease of the ambient temperature can reduce the magnitude of the HRM by 9–10% [7]. In parallel, it is reported that decrease of the ambient temperature by 1 °C corresponds, on average, to 2.1 less deaths/day per 100,000 people, or a drop in the initial average HRM per 1,000,000 people of 19% [7].

It is also reported that additional greenery in cities can considerably reduce the levels of heat-related morbidity. It is characteristic that in the city of Darwin,

Australia, increase of the GI by 20% can reduce annual excess hospital admissions by 40.1 to 27.5 per 100,000 people [36], while in Sydney, Australia, the addition of two million well-irrigated tress is found to decrease the excess heat-related morbidity from 3.7 hospital admissions/day to approximately 2.6/day per 100,000 people [37].

5 Conclusions

The climate of coastal cities is strongly affected by the presence of sea breeze that is their main cooling mechanism. Despite the presence of the sea breeze, coastal cities present considerable overheating problems affecting their environmental quality and economic activity. Overheating in coastal areas may reach 10 °C during the summer period, increasing considerably the energy consumption of buildings for cooling, the peak electricity demand, the levels of heat-related mortality and morbidity as well as the global environmental quality.

To counterbalance the impact of urban overheating in coastal cities, innovative mitigation technologies able to reduce the peak ambient temperature up to 4 °C have been developed and implemented. Post assessments have shown that the implementation of mitigation technologies reduce significantly the energy consumption for cooling purposes, while considerably decreasing the magnitude of health-related problems in cities.

References

1. Santamouris, M.: Recent progress on urban overheating and heat Island research. Integrated assessment of the energy, environmental, vulnerability and health impact. Synergies with the global climate change. Energ. Buildings. **207**, 109482 (2020)
2. Santamouris, M.: Analyzing the heat Island magnitude and characteristics in one hundred Asian and Australian cities and regions. Sci. Total Environ. **512–513**, 582–598 (2015)
3. Akbari, H., Cartalis, C., Kolokotsa, D., Muscio, A., Pisello, A.L., Rossi, F., Santamouris, M., Synnefa, A., Wong, N.H., Zinzi, M.: Local climate change and urban Heat Island mitigation techniques – the state of the art. J. Civ. Eng. Manag. **22**(1), 1–16 (2016)
4. Santamouris, M., Ding, L., Fiorito, F., Oldfield, P., Paul Osmond, R., Paolini, D., Prasad, A.: Synnefa: passive and active cooling for the outdoor built environment – analysis and assessment of the cooling potential of mitigation technologies using performance data from 220 large scale projects. Sol. Energy. **154**, 14–33 (2017)
5. Khan, A., Khorat, S., Doan, Q.-V., Khatun, R., Das, D., Hamdi, R., Carlosena, L., Santamouris, M., Georgescu, M., Niyogi, D.: Exploring the meteorological impacts of surface and rooftop heat mitigation strategies over a tropical city. J. Geophys. Res. Atmos. **128**, e2022JD038099 (2023). https://doi.org/10.1029/2022JD038099
6. Santamouris, M., Paolini, R., Haddad, S., Synnefa, A., Garshasbi, S., Hatvani-Kovacs, G., Gobakis, K., Yenneti, K., Vasilakopoulou, K., Feng, J., Gao, K.: Heat Mitigation Technologies can improve sustainability in cities. A holistic experimental and numerical impact assessment of urban overheating and related heat mitigation strategies on energy consumption indoor

comfort vulnerability and heat related mortality and morbidity in cities. Energ. Buildings. **217**, 110002 (2020)

7. Santamouris, M., Fiorito, F.: On the impact of modified urban albedo on ambient temperature and heat related mortality. Sol. Energy. **216**, 493–507 (2021)
8. Mat, S., Osmond, P.: Increasing green infrastructure in cities-impact on ambient temperature, air quality and heat related mortality and morbidity. Buildings. **10**, 233 (2020). https://doi.org/10.3390/buildings10120233
9. Haddad, S., Paolini, R., Synnefa, A., De Torres, L., Prasad, D., Santamouris, M.: Integrated assessment of the extreme climatic conditions, thermal performance, vulnerability, and well-being in low-income housing in the subtropical climate of Australia. Energ. Buildings. **272**, 112349 (2022)
10. Freitas, E., Rozoff, C.M., Cotton, W.R., Dias, S., Pedro, L.: Interactions of an urban heat island and sea-breeze circulations during winter over the metropolitan area of São Paulo, Brazil. Bound. Layer Meteorol. **122**, 45 (2007)
11. Ado, H.Y.: Numerical study of the daytime urban effect and its interaction with the sea-breeze. J. Appl. Meteorol. **31**, 1146–1164 (1992)
12. Gedzelman, S.D., Austin, S., Cermak, R., Stefano, N., Partridge, S., Quesenberry, S., Robinson, D.A.: Mesoscale aspects of the urban heat island around New York City. Theor. Appl. Climatol. **75**, 29–42 (2003)
13. Sakakibara, Y., Emi, M.: Relation between heat island intensity and city size indices/urban canopy characteristics in settlements of Nagano Basin, Japan. Geogr. Rev. Jpn. **78**(12), 812–824 (2005)
14. Lin, C.-Y., Fei, C., Huang, J.C., Chen, W.-C., Liou, Y.-A., Chen, W.-N., Liu, S.-C.: Urban heat island effect and its impact on boundary layer development and land–sea circulation over northern Taiwan. Atmos. Environ. **42**, 5635–5649 (2008)
15. Ohashi, Y., Kida, H.: Local circulations developed in the vicinity of both coastal and inland urban areas: a numerical study with a mesoscale atmospheric model. J. Appl. Meteorol. **41**, 30–45 (2002)
16. Khan, S.M., Simpson, R.W.: Effect of a heat island on the meteorology of complex urban airshed. Bound. Layer Meteorol. **100**, 487–506 (2001)
17. Yoshikado, H.: Interaction of the sea-breeze with urban heat islands of different sizes and locations. J. Meteorol. Soc. Jpn. **72**, 139–143 (1994)
18. Sakaida, K., Egoshi, A., Kuramochi, M.: Effects of sea breezes on mitigating urban heat island phenomenon: vertical observation results in the urban center of Sendai. J. Geogr. **120**, 2382–2391 (2011)
19. Hua, L.J., Ma, Z.G., Guo, W.D.: The impact of urbanization on air temperature across China. Theor. Appl. Climatol. **93**, 179–194 (2008)
20. Sadiq, N., Ahmad, I.: An empirical microclimate analysis of the coastal urban city and comparison of vicinal urban and rural areas with it. Pak. J. Meteorol. **6**, 11 (2010)
21. Devi, S.S.: Urban heat islands and environmental impact. Sixth Symposium on the Urban Environment AMS Forum: Managing our Physical and Natural Resources: Successes and Challenges (Tuesday, 31 January 2006)
22. Santamouris, M., Synnefa, A., Karlessi, T.: Using advanced cool materials in the urban built environment to mitigate heat islands and improve thermal comfort conditions. Sol. Energy. **85**, 3085–3102 (2011)
23. Synnefa, A., Santamouris, M., I.: Livada: a study of the thermal performance of reflective coatings for the urban environment. Sol. Energy. **80**(8), 968–981 (2006) Solar Energy during 2005–2006
24. Synnefa, A., Santamouris, M., Apostolakis, K.: On the development, optical properties and thermal performance of cool colored coatings for the urban environment. Sol. Energy. **81**, 488–497 (2007)
25. Garshasbi, S., Feng, J., Paolini, R., Duverge, J.J., Bartesaghi-Koc, C., Arasteh, S., Khan, A., Santamouris, M.: On the energy impact of cool roofs in Australia. Energy Build. **278**, 112577 (2023)

26. Garshasbi, S., Santamouris, M.: Using advanced thermochromic technologies in the built environment. Recent development and potential to decrease the energy consumption and fight urban overheating. Sol. Energy Mater. Sol. Cells. **191**, 21–32 (2019)
27. Karlessi, T., Santamouris, M., Apostolakis, K., Synnefa, A., Livada, I.: Development and testing of thermochromic coatings for buildings and urban structures. Sol. Energy. **83**(4), 538–551 (2009)
28. Santamouris, M., Feng, J.: Recent progress in daytime radiative cooling: is it the air conditioner of the future? Buildings. **8**(12), 168 (2018). https://doi.org/10.3390/buildings8120168
29. Feng, J., Khan, A., Santamouris, M.: The Heat Mitigation Potential and Climatic Impact of Super-Cool Broadband Radiative Coolers on a City Scale. Cell Reports Physical Science. **2**, 100485 (2021)
30. Khan, A., Carlosena, L., Feng, J., Khorat, S., Khatun, R., Doan, Q.-V., Santamouris, M.: Optically modulated passive broadband daytime radiative cooling materials can cool cities in summer and heat cities in winter. Sustainability. **14**, 1110 (2022). https://doi.org/10.3390/su14031110
31. Gao, K., Santamouris, M.: The use of water irrigation to mitigate ambient overheating in the built environment: recent progress. Build. Environ. **164**, 106346 (2019)
32. Santamouris, M., Ban-Weiss, G., Cartalis, C., Crank, C., Kolokotsa, D., Morakinyo, T.E., Muscio, A., Edward Ng, P., Osmond, R., Paolini, A.L., Pisello, F., Rossi, D., Sailor, A., Synnefa, H., Taha, H., Takebayashi, M., Zinzi, J.Z., Tan, Z.: Progress in urban greenery mitigation science – assessment methodologies advanced technologies and impact on cities. J. Civil Eng. Manag. **24**(8), 638–671 (2018)
33. Adilkhanova, I., Santamouris, M., Yun, G.Y.: Coupling urban climate modeling and city-scale building energy simulations with the statistical analysis: climate and energy implications of high albedo materials in Seoul. Energ. Buildings. **290**, 113092 (2023)
34. Garshasbi, S., Feng, J., Paolini, R., Duverge, J.J., Bartesaghi-Koc, C., Arasteh, S., Khan, A., Santamouris, M.: On the energy impact of cool roofs in Australia. Energ. Buildings. **278**, 112577 (2023)
35. Salamanca, F., Georgescu, M., Mahalov, A., Moustaoui, M., Martilli, A.: Citywide impacts of cool roof and rooftop solar photovoltaic deployment on near-surface air temperature and cooling energy demand. Bound.-Layer Meteorol. **161**(1), 203–221 (2016)
36. Haddad, S., Paolini, R., Ulpiani, G., Synnefa, A., Hatvani-Kovacs, G., Garshasbi, S., Fox, J., Vasilakopoulou, K., Neild, L., Santamouris, M.: Holistic approach towards urban sustainability: co-benefits of urban heat mitigation in a hot humid region of Australia. Sci. Rep. **10**, 14216 (2020)
37. Garshasbi, S., Haddad, S., Paolini, R., Santamouris, M., Papangelis, G., et al.: Urban mitigation and building adaptation to minimize future cooling energy needs. Sol. Energy. **204**, 708–719 (2020)

Sick Ships: A Discussion on Historical Cases and Optimization for the Future

Stamatina Th. Rassia and Thodoris Emm Tsikis

1 Introduction

1.1 A Brief Historical Overview

Famous cases of deadly ships have marked history. Some became legendary over the period of the black plague and the Spanish flue. Certain characteristic cases are the following.

- *Ghost ship run aground in 1349 in West Coast of Norway*, near the harbor of Bergen [1], bearing its wool cargo and dead crew. This ship had departed from England and was sailing north. All crewmembers suffered from the bubonic plague and died onboard bringing the ship into an aimless drifting with its wool cargo and its deceased crew. The legend says that only fleas and rats were alive on board, which brought the plague to the Scandinavian nation.
- *Grant Saint-Antoine Three-masted French ship quarantine in 1720, in the Southern port of France, Marseille* [2]. This ship, which carried precious silk fabrics departed from Lebanon with nine passengers who suffered from the plague. According to the quarantine protocols in Marseille, the ship, crew, passengers, and cargo had to be held in quarantine isolation in a nearby island. The silk merchants requested to receive earlier the precious cargo. Protocols were breached and the cargo was taken out of the ship earlier than advised. A month later a woman died of plague and soon after 100 people per day. Wealthy citizens and governmental authorities fled. Prisoners were forced to carry dead bodies

S. Th. Rassia (✉)
Le.D.R.A. Group, Athens, Greece
e-mail: contact@ledragroup.org

T. E. Tsikis
Tsikis Shipyard, Perama, Greece
e-mail: info@tsikisboatyard.com

S. Th. Rassia (ed.), *The Blue Book*, https://doi.org/10.1007/978-3-031-48831-3_7

from the streets. In 1721 King Louis XV enforced quarantine using armed forces over the whole area.

- *RMS Niagara ocean liner in 1918 Spanish flu case* [3]. The ship arrived from Vancouver to Auckland with reportedly two cases of "simple" influenza onboard. As she carried onboard William Massey who was the Prime Minister of New Zealand at the time and his Finance Minister Sir J. Ward, quarantine was refused in order to avoid inconveniencing the statesmen. RMS Niagara sailed to Honolulu where tens of crewmembers and passengers were falling ill. Second class cabins were converted into temporary hospitalization wards. One person died and many who were contaminated suffered from pneumonia, which was a common threat of the Spanish flu [4]. This ship was associated with the second wave of the Spanish flu in New Zealand [5, 6].

Quarantine (or Quarantina) has been considered as a way of mitigating the spreading of the Plague over the fourteenth century. The word was derived from the Italian "quaranta giordi," which meant "forty days" and referred to the period of isolation of all ships and people before they could dock in the city of Dubrovnik (in Croatia) at the time Croatia was under Venetian Rule [7]. This protocol was associated with the Plague that spread fast across Europe wiping out about half of its population from 1345 to 1360 [7]. A few hundred years later, in 1894, the Swiss-French bacteriologist and physician Alexandre Yersin (1863–1943) who worked at the Pasteur Institute discovered the causation of the plague while analyzing an outbreak in Hong Kong at the time. This bacterium was named after him as *Yersinia pestis* and was originating from fleas in Asia that infested mice and rats. The latter were frequenting ships and seemed to have spread the disease across the globe [7].

2 Safety and Health Onboard Ship

One of the most crucial aspects of safety at sea, is the preservation of good health. Should health ailments, injury, communicable diseases, or other life threating conditions occur, the International Maritime Organization (IMO), International Labour Organization (ILO) Maritime Labour Convention, World Health Organization (WHO), supported by the International Transport Workers' Federation (ITF), and International Shipping Federation (ISF) have composed more than 30 chapters with groups of incidents that can be managed onboard ships. These have been presented in the "International Medical Guide 3rd Edition" [8] and have provided essential information for health and safety while sailing onboard vessels.

There are two (2) common terms to describe infected person(s) onboard that are mainly used to classify the condition of infection and potential of transmission. These are, for instance, that a person is a (i) "carrier," of an illness that can be passed on to others, and (ii) "contact" of the infected person and therefore might also contract the disease [8].

Disease typologies could be crudely categorized as infectious and noninfectious. For instance, if an illness is endogenic, e.g., a heart attack, then this will not be treated as a public health outbreak but as a case-specific one. If the disease however is considered as an infectious disease, then there are three important periods of time in order to take care of the patient and at the same time mitigate further spreading. These are (i) isolation, (ii) segregation, and (iii) quarantine periods. Each of these periods refers to different conditions. Thus, the isolation period refers to ill patients' isolation to a specific place in order to mitigate spreading and receive care. The segregation period applies to a person potentially incubating a virus. The quarantine refers to protocols under which port authorities can or cannot admit a contaminated ship to shore [8]. The latter is known to be effectuated in cases of illnesses, such as these of the plague, yellow fever, Spanish fever, cholera, as well as recently COVID-19.

Suffering a headache, high fever, experiencing "lack of energy," muscular pains, and more can be symptoms of the onset of a communicable/infectious disease but might not be identified as such, until very clear symptoms of a specific disease have been diagnosed. Flu-like symptoms might last for a little while over the course of a day, or a few days. Therefore, there are numerous outlines of precautions and prescribed methods of management should they are identified on board, as stated by WHO. Workers onboard ships, spend a significant part of their lives onboard vessels, hence their injuries and illnesses are not to be taken lightly. When a vessel carries goods or passengers, fuels, and cargo then a disease outbreak onboard can cause a multitude of complications.

2.1 *Brief Case on Cruise Ships*

Understanding the great value of the prevention, management, and mitigation of communicable disease outbreaks on board is also the topic of the EU-funded project Healthy Ship 4 U (HS4U, GA number: 101069937), which focuses on the above-mentioned for the case of modern cruise ships. Thus, a variety of experts and hands-on experience from different disciplines of research and practice are offered for the development of viable solutions that can be implemented in the future of cruise ship sailing. Remarkably, such ships might operate transporting 8000 passengers and crew at a time, which is a substantial number of people.

Environmental control methods to preserve indoor hygiene [8] onboard ships include: disinfection techniques; daylight conditions; ventilation conditions; crew services; food handling and storage; toilets hygiene; solid waste management; kitchen area hygiene; potable water systems and sanitation inspection; as well as the combating of numerous potential disease vectors.

The IMO has announced "safety and environmental standards on passenger ships" [9]. The Safety of Life at Sea (SOLAS) treaty, in addition, has been applied for international passenger voyages, when the number of passengers is larger than 12 people. Treating sewage systems [10] of cruise ships has been one of the pollution preventive regulations in place by IMO MARPOL, which can be coupled with multiple Annexes on relevant health protection measures for people in ports, coastal areas, as well as the passengers and crew onboard ships. These Annexes target energy efficiency, air handling, and multilevel pollution requirements.

Crewmembers receive training for specific codes of conduct when working on passenger ships so that they are prepared to face health and safety challenges as well as different emergency situations. Crisis management is heavily related to human behavior onboard; therefore the crew is responsible for numerous safety protocols that must be applied in order to ensure safe return to the port [9].

3 Blue Growth Strategy and the Value of Innovation

Aiming to support and promote smart and sustainable methods to induce the EU goals for growth in the marine and maritime sectors, an ecosystem of services [11] for shipping should be analyzed. Blue growth strategy in coastal areas as well as in shipping could, for instance, provide successful resolution of crises when receiving contaminated ships. This could help, by design and legislation, to decrease the negative effects of ship quarantine and the related impacts over local and global economies as well as social and environmental sustainability.

The century long history of shipping is closely related to coastal areas and related growth. Supporting the maritime sector to face risks of infectious diseases could improve coastal and global economies. There are hundreds of thousands of jobs related to the blue economy [11], hence, optimizing conditions of operation within the motorways of the sea could improve digital medical reporting and minimize hazard mitigation through design innovation.

Relevant key enablers of change therefore could design methods of administrating complexities with respect to health crises onboard and onward on land. Improving livelihood and an ecosystem of health can further proactively follow the World Bank definition that:

> The blue economy is sustainable use of ocean resources for economic growth, improved livelihoods and jobs, and ocean ecosystem health [12].

4 Conclusion

While it has been identified that numerous actions, guidelines, protocols, and regulations have been set up internationally, the optimization of disease prevention still constitutes an open question of intense research activity. Future design requirements will thus be explored over the course of a number of projects including the EU-funded project Healthy Ship 4 U (HS4U, GA number: 101069937).

References

1. Verb Company: Seven centuries later, another ghost ship is coming to Norway. https://verbcompany.com/2017/07/28/seven-centuries-later-another-ghost-ship-coming-norway/ (2017). Accessed 26/01/2023.
2. De Rosen, F.: Stories of Space in Times of Quarantine. McHarg Center for Urbanism and Ecology, University of Pennsylvania Stuart Weitzman School of Design (2020). https://mcharg.upenn.edu/stories-space-times-quarantine
3. Denniston, J.E., Mitchelson, E., McLaren, D.: Influenza Epidemic Commission (report of the). Marcus F Marks – via AtoJs online, Wellington (1919)
4. Patients from the Niagara. The Colonist. 17 October 1918. p. 7. Retrieved 19 December 2020 – via Papers Past
5. Rice, G.: Black November: The 1918 Influenza Pandemic in New Zealand, revised edn, pp. 198–199. Canterbury University Press, Christchurch (2005)
6. https://en.wikipedia.org/wiki/RMS_Niagara
7. Schwarcz, J.: The Word "Quarantine" Comes from the Italian Word "Forty Days". McGill Office of Science and Society. https://www.mcgill.ca/oss/article/did-you-know-health/word-quarantine-comes-italian-word-forty-days (2020). Accessed 27/01/2023
8. World Health Organization: International Medical Guide for Ships, 3rd Edition: Including the Ship's Medicine Chest. Switzerland. https://apps.who.int/iris/bitstream/handle/10665/43814/9789240682313_eng.pdf?sequence=1&isAllowed=y (2007). Accessed 08/02/2023
9. International Maritime organization (IMO): Safety and environmental standards on passenger ships. https://www.imo.org/en/MediaCentre/HotTopics/Pages/passengership-default.aspx (2019). Accessed 08/02/2023
10. IMO MARPOL: Prevention of pollution by sewage from ships. https://www.imo.org/en/OurWork/Environment/Pages/Sewage-Default.aspx. (2019). Accessed 08/02/2023
11. European Parliament Research Service, Lead author: Scholaert F. with contributions from Margaras, V., Pape, M., Wilson, A., Kloecker, C.-A.: The blue economy: overview and EU policy framework. https://www.europarl.europa.eu/RegData/etudes/IDAN/2020/646152/EPRS_IDA(2020)646152_EN.pdf (2020). Accessed 08/02/2023
12. Blue Economy Definitions. https://www.un.org/regularprocess/sites/www.un.org.regularprocess/files/rok_part_2.pdf. Accessed 08/02/2023

Disease Spread Control in Cruise Ships: Monitoring, Simulation, and Decision Making

Georgios Triantafyllou, Panagiotis G. Kalozoumis, Eirini Cholopoulou, and Dimitris K. Iakovidis

1 Introduction

Cruise ships are closed environments that combine private and public areas, where diverse populations travel together for several days. Such conditions are considered ideal for the spread of both airborne and waterborne diseases. A recent example is the Diamond Princess cruise ship, where after a passenger was tested positive for COVID-19 (SARS-CoV-2), the passengers and the crew were instructed to quarantine for 2 weeks. During that period, there was an extensive spread of the virus throughout the ship, counting 651 cases over 3711 passengers and crew [1]. After almost 1 month, all passengers and crew disembarked. Notably, 712 (19.2%) were tested positive for COVID-19, and among them, 331 (46.5%) were found asymptomatic [2].

In general, disease outbreaks are relatively common in closed environments, such as cruise ships, nursing homes, hospitals, and college dormitories [3]. In the case of cruise ships, outbreaks are associated with either airborne diseases, e.g., COVID-19 or waterborne diseases, e.g., gastrointestinal (GI) diseases. These are usually attributed to organisms, such as Salmonella, Shigella, enterotoxigenic *Escherichia coli*, influenza virus, and *Legionella pneumophila*, and, more recently, norovirus and COVID-19 [4, 5]. During such outbreaks, cruises often go into quarantine until the source is isolated and the spread is mitigated. Therefore, modeling and predicting the progress of disease outbreak in closed environments is of paramount importance for their mitigation. The success of a modeling approach depends highly on the parameters taken into consideration, such as social and behavioral factors, as well as parameters related to the transmission pathways. For the latter, accurate

G. Triantafyllou · P. G. Kalozoumis · E. Cholopoulou · D. K. Iakovidis (✉)
Department of Computer Science and Biomedical Informatics, University of Thessaly, Lamia, Greece
e-mail: gtriantafyllou@uth.gr; pkalozoumis@uth.gr; echolopoulou@uth.gr; diakovidis@uth.gr

S. Th. Rassia (ed.), *The Blue Book*, https://doi.org/10.1007/978-3-031-48831-3_8

data regarding the transmission through the air, water, and surfaces are key to the effectiveness and fidelity of a simulation.

Diseases can spread via different transmission modes, including airborne, droplet, contact, and fecal-oral transmissions [6]. The most common disease transmission pathway among individuals is air. An infectious disease, such as COVID-19, can be spread when contact is established between an infected host and a susceptible one. Peer-to-peer contact modes involve complex interactions of a pathogen with a fluid phase, e.g., isolated complex fluid droplets or a multiphase cloud of droplets. When an individual exhales, including coughing or sneezing, micron-sized droplets are formed, which can transfer airborne pathogens, such as such as viruses ($\sim$10–100 nm), bacteria ($\sim$1 μm), and spores ($\sim$1–10 μm) [7]. Fine and ultrafine particles (airborne transmission) can stay suspended for an extended period of time ($\geq$2 h) and be transported through simple diffusion and convection mechanisms as far as 8 m [6]. These can be either transmitted directly to other people or be deposited on surfaces and transmitted indirectly through contact.

Another but less frequent transmission pathway is water. A direct source of infection is infected potable water, while premise plumbing with non-potable water, e.g., showers, toilets, and sink faucets, and the water systems associated with it, are common indirect sources of infection transmission [8]. Nevertheless, nowadays extensive measures are taken to minimize or even eliminate the occurrence of such cases [9, 10]. Air bubbles are ubiquitous on water surfaces and are common in facilities of cruise ships, such as pools and toilets. Bubbles can also cause public health concerns, since droplets generated from contaminated water may carry infectious payloads that can lead to airborne disease transmission. Contaminated droplets can be easily transported over short and long distances by drafts, such as those generated by the heating ventilation and air-conditioning (HVAC) system. Inhalation of droplets or droplet nuclei containing bacteria is a recognized route of infectious disease transmission, e.g., outbreaks of Legionella [11], which are attributed to aerosolization from bursting bubbles. Another source for the water-borne transmission of diseases is the development of biofilm in the piping systems, especially those with warm water (<50 °C). Biofilms develop on most surfaces in contact with non-sterile water. Pathogens in biofilms do not usually spread from person-to-person or through direct consumption of the water, but there is considerable evidence that individuals become ill when exposed to airborne water droplets that have been seeded by those pathogens. Activities that can lead to their aspiration include showering and hand washing.

Although viruses, such as SARS-CoV-2, are principally transmitted through person-to-person contact through droplets produced while talking, coughing, or sneezing, transmission may also occur through other routes, such as contaminated surfaces. However, the role that surfaces play on the spread of the disease remains contested. For example, infective coronavirus has been found to persist on surfaces from 3 h to 28 days, depending on different environmental factors, e.g., surface material, humidity, and temperature. Viruses can be transferred from contaminated surfaces to the hand upon contact and from the hand to the mucous membranes on the face. Nevertheless, experimental findings reported in the literature support the current perception that contaminated surfaces are not a primary mode of transmission, at least as regards SARS-CoV-2 [12, 13].

Monitoring and controlling the spread of a disease in a cruise ship is a complex problem, requiring information about the ship (e.g., accessible areas, ventilation, and water supply network), the passengers (e.g., demographic and health data), and possible diseases, including their symptoms and treatments, the respective pathogens and epidemiology. To this end, several guidelines and protocols have been developed [14, 15]; however, their application is mainly based on decisions and interventions performed by the vessel's crew. Therefore, managing a crisis, such as an epidemic, on board is currently prone to human errors, which can be life-threatening in the case of severe infectious diseases. The frequency of such errors depends on several parameters, including the number of the crew members, their training and experience, the size of the vessel, and the number of passengers on board. Furthermore, the interaction of crew members with diseased passengers can contribute to further spreading of the disease and affect its operational capacity.

It is therefore evident that an automatic system for disease spread monitoring and control in cruise ships would contribute to limiting the dependencies on the human factor, and consequently to passengers' safety. Toward this direction, this chapter presents a conceptual model integrating technologies that could be used for disease monitoring and spread control. This model integrates sensors, actuators, artificial intelligence, mathematical, numerical, and/or other simulation models to predict the risk of disease spread in cruise ships, and to support decision making regarding possible mitigation measures.

The remainder of this chapter is organized as follows: Sect. 2 introduces a conceptual closed-loop model for disease spread control in cruise ships; Sect. 3 summarizes diseases monitoring approaches and equipment focusing on ships and enclosed spaces; Sect. 4 conducts an extensive review on epidemic-related risk assessment (RA) methodologies, and provides a series of risk management measures for outbreaks on ships; Sect. 5 provides a literature review on key studies related to airborne, waterborne, and surface/fomite-mediated transmission modeling; Sect. 6 summarizes artificial intelligence (AI)-based reduced-order models that can be utilized to reduce the computational complexity of the models in Sect. 4; Sect. 7 summarizes the research challenges identified after reviewing the relevant literature; and Sect. 8 summarizes a set of conclusions derived from this study.

2 A Conceptual Closed-Loop Model for Disease Spread Control

In the context of disease spread monitoring various sensors, such as thermal cameras, can be used to derive semantic information about the passengers and the condition of the different areas of the ship after proper analysis of the raw data they acquire, e.g., identify disease symptoms, such as fever, or identify increased passenger concentrations in public areas. The current guidelines and protocols, as well as the experience from multiple disciplines involved, such as experience from medical and naval personnel, can be integrated, e.g., as rules, into knowledge-based decision-making systems, to automatically infer useful information for disease control. Such information may include the infection risk in different areas of the ship over time, and possible actions that need to be taken to limit the spread of the disease. Furthermore, simulation models can be used to predict the evolution of the disease spread considering both microscopic parameters, such as the physical properties of the pathogens and their molecular interactions, and macroscopic parameters, such as the motility of the passengers on board.

A cruise ship may include multiple decks with different spaces accessible by the passengers, e.g., cabins, cabin corridors, lounges with sitting areas, shops, and outdoor spaces, such as canteens and swimming pool areas. The risk for the spread of an infectious disease in these spaces may vary, e.g., bathrooms and restrooms are likely to exhibit a higher risk than free spaces; therefore, closer monitoring with special sensors may be required.

Figure 1 provides a schematic representation of the architecture of a generic RA model (GRAM). A fundamental component of this architecture is the RA node (RAN) (Fig. 1a). GRAM is fully modular, composed of multiple interconnected RANs, set up for local monitoring and decision making in different areas within a cruise ship, as illustrated in Fig. 1b. A RAN ρ considers an environment e in a ship area of interest, monitored by a set of sensors. The raw data from the sensors, along with relevant data that may be available from the information system of the ship, are processed and analyzed to recognize and extract semantic information about the status of the environment and possible events taking place in that environment, e.g., using a machine learning (ML)-based classification system. This higher-level information is provided as input into a decision-making algorithm designed to infer the risk r_ρ of the environment being contaminated with a pathogen, such as SARS-CoV-2, and possible actions a_ρ to mitigate this risk. Implementing the inferred actions depends on the confidence of inferred decisions. If the confidence is low, implying a high uncertainty based on the available data, a simulation module is activated. The simulation module will provide RAN with additional data that could enhance the confidence of the decision making. Running simulations will enable testing various alternative scenarios in silico, e.g., using different passenger populations and activities, or predicting the outcome of different mitigation measures. Thus, based on the predicted data, the decision-making module should be able to infer decisions of higher confidence about the risk of disease spreading over time. It is important to consider the risk as a function of time, $\overline{r}_\rho(t)$, for future values of t, because the spread of a disease depends on the evolution of pathogens both in space and time. Also, the decision-making module will also be able to infer the actions for risk mitigation with higher confidence. These actions will result in changes in the monitored environment e, and the whole monitoring and disease control process will follow a closed-loop approach (where the decision-making module plays the role of the controller), like the one typically followed in automatic control systems and robotics [16].

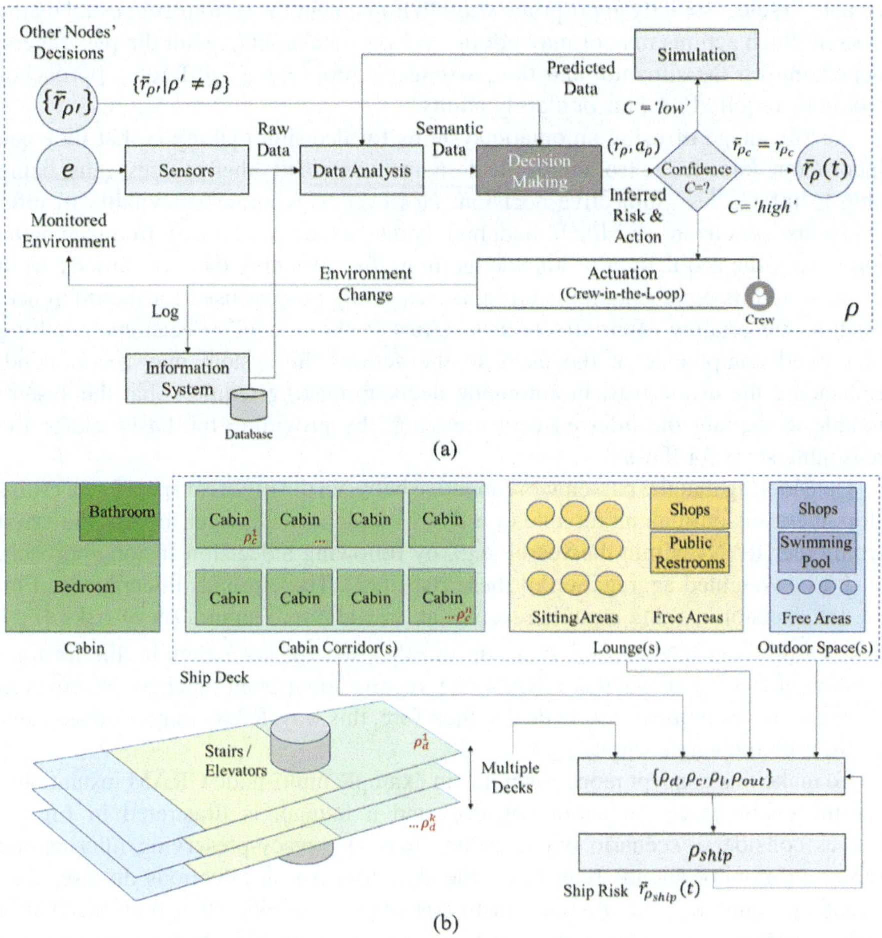

Fig. 1 Generic closed-loop model for disease control in passenger ships. (**a**) Risk assessment node. (**b**) Multi-node risk assessment

The actions needed to effectively mitigate the risk of the spread of an infectious disease, should consider social aspects, since a cruise ship is usually a diverse human community that includes people with different personalities, ages, cultural background, and habits. Also, passengers are customers investing in their cruise vacation, and despite the critical situation that must be managed by strictly following relevant health protocols, they should feel comfort and convenience within a safe, caring environment. To this end this chapter introduces the concept of the *Crew-in-the-Loop* (CiL). Instead of having a fully autonomous actuation system for disease spread control, the CiL approach considers that the crew is continuously being informed by the decision-making system about the risk in different spaces of the ship and a set of possible mitigation actions that can be taken. Thus, the crew will be able to select socially appropriate ways to implement the actions suggested by the system. Such actions may or may not involve personal contact with the passengers, depending on the situation and the passengers' profile, e.g., different approaches would be required for kids or elderly adults.

An advantage of using automatic systems for decision making is that they can incorporate knowledge from multiple domain experts and other sources; thus, being able to infer a less subjective decision. However, it is almost inevitable to infer decisions without an error likelihood, mainly due to the uncertainty introduced in the input data, the acquisition of knowledge (e.g., from training data, or directly from experts) and its representation, and in the reasoning process used for the inference. Such an uncertainty often affects users' trust in the system's decisions, resulting in limited compliance of the users to the actions the system may recommend. Enhancing the users' trust in automatic decision making requires that the system is able to explain the inferred decisions, e.g., by providing the basis and/or the reasoning steps for these decisions.

Considering that the passengers can move between the different spaces of a cruise ship, the decision-making module of a RAN should also consider the risks inferred from other RANs within the vessel, e.g., by following a decision-fusion approach, such as a weighted aggregation of these risks [17]. Therefore, as illustrated in Fig. 1a, the decision-making module receives as an additional input a set of risks $\{\overline{r}_{\rho'}\}$ originating from RANs $\rho' \neq \rho$ set up in other spaces. For example, the abstract scheme of Fig. 1b shows that a RAN may receive input from other RAN setups in different spaces of the different decks; therefore, this way, RANs may address wider areas, even the whole ship (ρ_{ship}).

To make this concept more concrete, an example multi-node GRAM instantiated for the cabin spaces, a cabin corridor, and a lounge, is illustrated in Fig. 2. Let us consider a scenario where cabins have a privacy-preserving microphone sensor to monitor sounds relevant to the symptoms of an infectious disease, e.g., sneezing, coughing, etc., or relevant to passengers' activities that may result in a contamination, e.g., visiting the toilet. As illustrated in Fig. 2a the data analysis module of the cabin's RAN receives as input the sounds captured by the microphone but also relevant data about the passenger from the information system of the ship. The semantic data resulting from the data analysis are subsequently inputted into the decision-making module. Assuming that the inferred decision about the risk of the space being infected by a pathogen, e.g., SARS-CoV-2, Norovirus etc., has a sufficiently high confidence level, the system informs the crew to take safety measures, e.g., the ship's medical doctor (MD) to examine the situation and to provide masks and sanitizers to the passengers in the cabin if necessary.

Considering that all cabins in a corridor of the ship are equipped with the same RANs, a RAN dedicated to the corridor area can be set up. Fig. 2b illustrates such a RAN, in a corridor area without any sensors. The data analysis module of this RAN processes relevant data from the ship's information system, e.g., if the cruise ship has stopped in a destination or not, embarkation and disembarkation data, and the processed data are entered into the decision-making module, which co-evaluates the

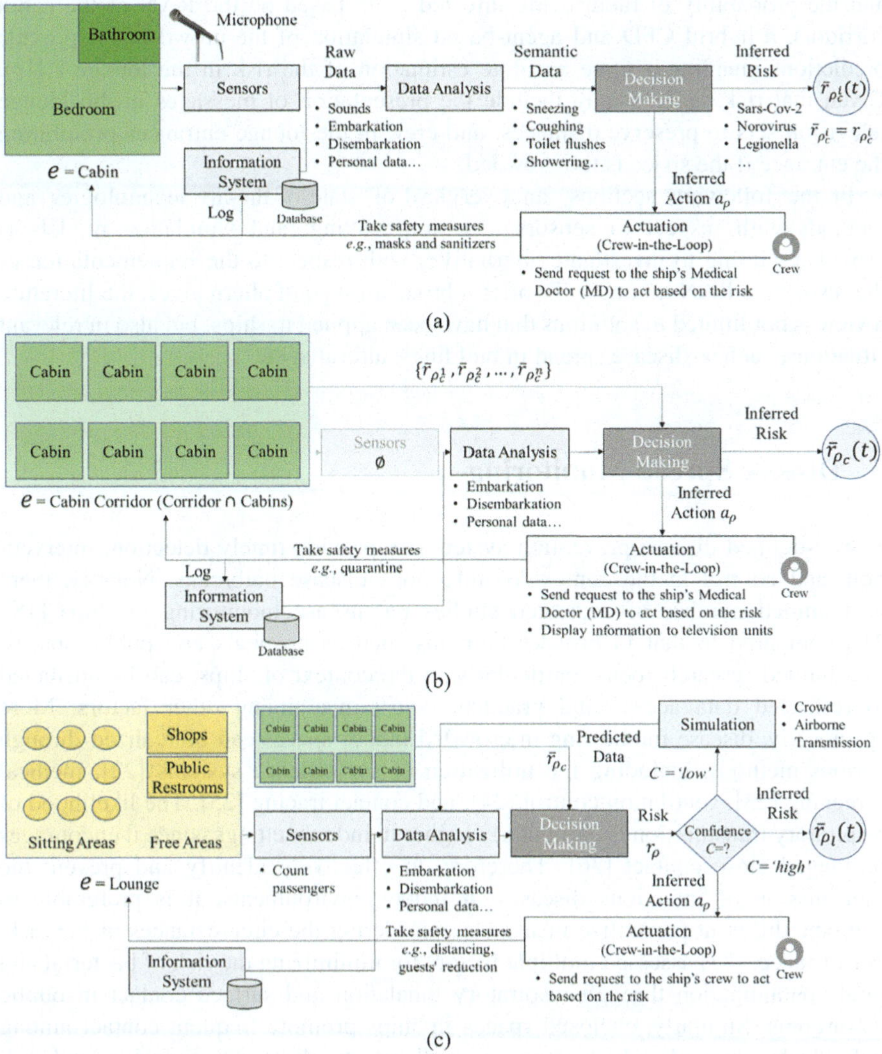

Fig. 2 Example implementations of the closed-loop model for different ship spaces. (**a**) Cabin. (**b**) Cabin corridor. (**c**) Lounge

risks inferred from the RANs of all other cabins in the corridor. Similarly, the risk of a lounge area is inferred by considering the risk inferred from the cabin corridor that resides in the same deck. Fig. 2c illustrates such an example, where the lounge has a sensor to count the passengers entering and/or leaving the area. Also, in this case, it is assumed that the confidence of the output risk is initially low; therefore, a simulation should be run to produce additional (predicted) data to increase the confidence. Considering the structure of the lounge space, the passengers' crowd, and the probability of them being infected (e.g., based on the RAN of the cabin corridor), a hybrid CFD and agent-based simulation of the crowd could provide predictions enabling a more accurate estimation of the risk in the lounge $\overline{r}_{\rho_l}(t)$. Given that risk, actions may include the presentation of messages in the lounge television sets to preserve distances, and crew in the lounge entrances prohibiting the entrance if the space is too crowded.

In the following sections, an overview of state-of-the-art technologies and methods with respect to sensors, decision making, and simulation models is provided, aiming to investigate alternatives with respect to the implementation of the respective RAN modules. To offer a broader range of alternatives, the literature review is not limited to solutions that have been applied to ships, but also in relevant situations, such as disease spread in buildings, aircrafts, etc.

3 Disease Spread Monitoring

A well-defined disease prevention system can provide timely detection, intervention, and control in the context of infectious disease outbreaks. Notably, there is a limited number of dedicated studies on disease monitoring in ships [18–21], compared to that in broader domains such as hospitals and public spaces. The limited research focus, particularly in the context of ships, can be attributed to restricted data access and practical constraints among other factors. More specifically, disease monitoring in crowded indoor spaces can be realized through various methods, including the utilization of surveillance systems [22], medical screening [23], ventilation control [24], and contact tracing [25]. The likelihood of respiratory transmission is particularly higher in indoor settings, since it encourages prolonged close contact [26]. Therefore, to effectively identify and prevent the transmission of infectious diseases in indoor environments, it is preferable to combine different preventive measures, considering the circumstances of the task. For instance, [27] assessed multiple factors for minimizing the risk of bacterial and viral contamination through respiratory inhalation and surface contact in public washrooms. Similarly, enclosed spaces in ships promote frequent contact among individuals onboard, which poses a challenge to disease monitoring and risk management in the event of an outbreak [28]. In particular, the recent COVID-19 outbreaks has created the need for mechanisms that promptly identify, isolate, and treat the affected individuals, considering that the main routes of transmission include respiratory droplet inhalation, as well as direct and indirect contact with the virus [29].

Several methods incorporate smart technologies to develop effective outbreak response plans with the aim of mitigating such infectious disease epidemics on ships [30], either through Internet of Things (IoT) sensors, drones, thermal cameras, or mobile applications. Ultimately, sensors provide real-time data on environmental conditions, human behavior, and other elements that can affect the spread of infectious diseases. There are various types of smart devices and IoT sensors that could be used in ships for disease monitoring [30, 31]. These include wearable sensors, air quality sensors, and thermals cameras. In particular, wearable sensors can be utilized by passengers and crew members to track their movement and monitor epidemic indicators such as body temperature, heart rate, and respiration rate. Hence, wearable sensors can enable extensive contact tracing and early symptom identification that allow for rapid isolation and treatment of infected individuals [32]. However, they may raise data confidentiality and security concerns as they require personal participation and compliance from individuals. In addition, air quality sensors can detect the presence of airborne particles, including viruses. In addition, their utilization provides measurements of air temperature, humidity, and ventilation rates, which impact the spread of infectious diseases [19]. Therefore, the incorporation of air quality sensors enhances the effectiveness of ventilation systems and locates areas with poor air circulation. Nevertheless, most of them lack the capacity to differentiate between various types of airborne contaminants. Furthermore, thermal cameras can be used to identify elevated body temperatures that could indicate the presence of fever, a symptom of several infectious diseases [33]. Moreover, thermal cameras are considered noninvasive and can effectively scan crowded places, such as boarding areas. However, they are ineffective at detecting asymptomatic carriers, who may not have fever. There are also approaches that employ other types of sensors to measure environmental parameters, such as infrared-based sensors. For example the approach described in [34] investigated the deployment of wireless sensor units to assess indoor air quality. This was accomplished by employing infrared sensors to measure humidity, CO_2, and temperature, as well as low-power wireless networking and geo-statistic methods for the prediction of holistic indoor maps. Another study designed IoT-based smart glasses to identify suspicious cases of COVID-19 among individuals [35]. Testing individuals using laboratory tests might be not available or can delay the detection of infectious diseases; thus, the use of other screening methods, such as biosensors could provide a reliable alternative [36]. Signal processing and analysis of cough sounds with the use of an AI framework was introduced as an alternate diagnostic method for COVID-19 [37], and more recently, a method that utilized cough, speech, and breath sound processing for diagnosing COVID-19 was proposed [38]. This innovative way of diagnosing respiratory diseases could be helpful in an epidemic outbreak aboard a ship; thus, such algorithms could be employed for onboard testing, alleviating the need for dedicated medical equipment and even aid with passive surveillance of the ship, i.e., detecting disease-classified sounds in the common areas.

Another alternative direction concerns methodologies that employ IoT-based solutions to acquire data from different sensors and realize disease monitoring. These approaches investigate the implementation of IoT-based indoor air quality (IAQ) monitoring systems that collect real-time data for effective control of the environmental parameters [39, 40]. More specifically, most methodologies in this context utilize sensors and IoT devices that usually harness cloud computing services to optimize ventilation systems with low energy consumption. Recently, various studies have explored an alternative direction that combines ML techniques and sensor-based systems. These approaches aim to detect abnormal patterns in air quality data and provide insights regarding the reason for these observed

Table 1 Sensors used for disease spread monitoring

Study	Sensors	Disease	Monitoring	Application
[18–21]	Air quality	COVID-19	CO2, illuminance, relative humidity, temperature	Ships
[39, 40, 42]	Air quality	Respiratory, cardiovascular, pulmonary	CO2, relative humidity, temperature, air pressure, illumination	Educational facilities, office, living room
[23, 32]	Wearable	Respiratory, chronic, cardiovascular	Blood pressure, heart rate, motion, sweat, body temperature	Clinical environment
[30, 33, 35, 45]	Thermal cameras	COVID-19	Body temperature, blood pressure, heart rate, cough quality	Hospitals, public spaces
[34]	Infrared	N/A	Humidity, CO2, temperature	Office
[37, 38]	Sound	COVID-19	Respiratory sounds	Hospital
[24, 27]	Others (dryer, fabric)	COVID-19	Bacterial CFU, environmental variables	Public washrooms, bus

irregularities [41, 42]. ML models have also been utilized to perform disease monitoring, diagnosis, and prediction based on collected health data using sensors, e.g., body temperature and heart rate, while using cloud-based services for real-time data transfer [43, 44]. In this context, there are various types of Internet-based epidemic intelligence systems that opt for different functionalities, such as social distancing detection, disease identification, and contact tracing [45]. Notably, the previously mentioned methodologies showcase the importance of implementing effective disease monitoring strategies that could be incorporated in indoor settings, particularly in the context of infectious disease outbreaks in ships.

However, the utilization of sensors in disease monitoring systems introduces several limitations including maintenance, cost, and data security concerns [43]. In particular, the collection and sharing of personal health information from cameras, wearables, and other sensors for disease monitoring may be subject to regulations related to privacy concerns. For instance, the General Data Protection Regulation (GDPR) establishes rules for the collection of personal data and states that individuals must be appropriately informed regarding their use [46]. Therefore, to address these concerns it is essential to implement clear policies and privacy-preserving protocols that ensure the confidentiality of the collected data, as well as to obtain informed consent from individuals prior to sharing their information.

Current studies indicate that processing and analysis of the signals acquired from different sensors can provide useful cues regarding the spread of infectious diseases in closed environments. Based on the semantics derived from such an analysis, decisions for RA and management can be made by decision support systems designed for this purpose.

Table 1 provides an overview of the previous works and the types of sensors utilized for disease spread monitoring. In the following section a review of such systems, methods, and models, is provided.

4 Decision Making for Risk Assessment and Management

RA is defined as the process of gathering, processing, and storing information in order to assign a level of risk to specific scenarios, such as an epidemic outbreak [47]. The RA cycle can be divided into three stages, i.e., risk identification, risk analysis, and risk evaluation [48]. RA is usually coupled with control measures and continuous evaluation of the risk [47]. RA tools have been employed in healthcare to assess and manage various scenarios, such as disease transmission, assessment of the potential risk for a patient to develop different diseases, or even management of hospital operations. From toxicology to epidemic outbreaks, RA tools have been proposed to mitigate potential hazardous situations, such as exposure to toxic chemicals [49], transmitted diseases [50], or predict the risk of developing high-morbidity diseases, e.g., cancer [51]. COVID-19 has greatly affected the healthcare sector [52, 53] and this is also apparent based on the influx of scientific research papers related to the disease.

The RA field has also been in the center of this trend as it is paramount for managing epidemic outbreaks [54, 55]. Epidemic outbreaks on cruise and cargo ships have been extensively analyzed in the relevant literature [56–61]; nonetheless, there is a limited number of suggested RA tools applied in such scenarios [62–64]. Regarding the broader spectrum of disease spread RA, the most common tools are based on three principal types of methods, i.e., knowledge-based, ML, and hybrid methods.

4.1 Knowledge-Based Methods

Knowledge-based methods are methods that contain information in the form of rules or relations that guide the system to decide the appropriate action. Experts' knowledge or experimental data can be analyzed to generate these rules. Statistical analysis and mathematical modeling of patient data combined with golden-standards are also employed to generate such rules and establish the knowledge base that the health practitioners should utilize to make informed decisions [65].

Regarding the broader spectrum of disease-spread-related RA methods, conventional knowledge-based methods, such as case-based-reasoning (CBR) or expert methods are employed in emergencies [48, 66]. Emergency protocols are usually created with the help of experts in the healthcare sector, often based on Delphi studies, to assess the risk during epidemic outbreaks [47, 67, 68]. The World Health Organization (WHO) has also established a RA tool based on disease characteristics and experts' opinions focused on influenza pandemic outbreaks [69]. In [70], the Delphi method was combined with a fuzzy analytical hierarchical process to estimate a COVID-19 risk index based on 19 weighted factors provided by experts. CBR methods are derived by extracting information from previous scenarios, i.e., previous pandemic outbreaks [66]. Despite the importance of utilizing information from past events to create informed RA tools, these methods lack flexibility in terms of adjustability to new pathogens and different conditions [71]. To this end, dynamic CBR methods have been proposed. For example, a dynamic CBR model was generated by analyzing disease emergency scenarios using a Markov model. There, a flow of eight stages was described for a pandemic outbreak scenario. Apart from the use of experts and previously acquired information, epidemic RA methods should also take into consideration the nature of the pathogens, e.g., virus or bacteria, that causes the disease, since it is essential for estimating the risk of a potential pandemic outbreak [72].

Mathematical modeling is extensively used in epidemiology, and thus, this subsection reviews methods that utilize knowledge-based risk factors when assessing risks of an epidemic outbreak. These methods focus on utilizing epidemiological spatiotemporal data [73], population demographics, and mathematical models to predict the course of an epidemic outbreak [74–77]. In [75], a rapid RA tool referred to as "EpiRisk" combined disease-specific and country-specific factors, i.e., economy, healthcare system status, and population demographics, to predict disease outbreaks. In detail, 18 factors were inferred from a review of the relevant literature and previous epidemic outbreak data. After the statistically redundant risk factors were removed from the system, the remaining 14 risk factors were used to infer the risk of an epidemic outbreak. These factors were calculated for every case study (country and disease) and summed up to calculate an epidemic outbreak risk score that was classified into four levels of risk, i.e., low, moderate, high, and extreme.

It should be noted that RA tools related to disease spreading differ from traditional healthcare-related RA methods, since they incorporate variables of transportation and transmission to their models, thus including a social aspect in the RA loop. This also indicates the need to understand the transmission mechanisms of a pathogen for accurately simulating the spread of the disease. In [74], a RA method was proposed, which utilized a stochastic susceptible-exposed-infectious-remove (SEIR) type model based on disease, population, and healthcare system characteristics. A "so-called" seeding (ST) and doubling time (DT) model has been implemented in [76] for early RA of an epidemic using only the number of infected cases as information, providing an alternative time-efficient RA tool. ST indicates the time interval between the day of the first confirmed case of infection and the day when the total number of cases reaches the seeding number (total number of cases required to cause and epidemic outbreak). DT indicates the time interval for the total number of cases to increase twofold. In this study, ST and DT were calculated based on the infected cases and then used to determine the risk of an epidemic outbreak. More specifically, the model classified cases into four risk levels, namely, low (long ST and DT), moderately low (long ST and DT), moderately high (long ST and short DT), and high (short ST and DT).

In the early stages of the COVID-19 pandemic, spatiotemporal data were used to create a population-flow-based risk model by utilizing the population flow from outbreak sources in Chinese prefectures near the Wuhan area, since it was found that these regions were highly correlated with increased risk of an outbreak [78]. Two exponential models were proposed to assess the risk of an outbreak. The first model was employed to predict infected cases (daily or cumulative) in each prefecture, considering data such as population outflow and size. The second exponential model was used to dynamically assess the risk of disease spread in each infected prefecture based on the infected population outflow from Wuhan. Moreover, the study evaluated the capacity of gravity models to calculate the number of infected cases and population flow based on the Wuhan incident. Gravity models can calculate the population flow volume between two geographical areas based on distance and population size.

A similar method combined a population distribution dataset and origin destination flow dataset with a gravity model that was used to model spatial interaction. In addition, the SEIR epidemic model was utilized to simulate disease spread for precise intervention policies in epidemic outbreaks [73]. Another method utilizing spatial data was implemented in [79] for the RA of disease outbreaks in Italy, where seven risk indicators were proposed by the authors, i.e., housing concentration, healthcare density, air pollution, average winter temperature, age of population, mobility index, and population size. Two functions were proposed to model the effect of the risk indicators in the RA method, namely, a linear model and a multiplicative model. The former combined all the risk indicators in one linear function and the latter considered the product of two linear functions each taking three of the six risk indicators into account. The coefficients of the models were inferred by least-square fitting statistical analysis.

Uncertainty is inherent in medical decision making [80, 81], which also applies in the RA of epidemic outbreaks [82]. Fuzzy logic has been proven helpful to combat this problem, since it exploits the ambiguity found in real life [83]. This is further highlighted in [84], where the fusion of fuzzy systems with edge computing is considered a promising methodology for RA of epidemic outbreaks. A RA method of analyzing risk factors with the use of an interval-valued q-rung orthopair fuzzy set (IVq-ROFS) evaluation method was implemented in [85]. In particular, a set of initial risk factors were considered based on the Delphi method and were subsequently represented as IVq-ROFSs with weights corresponding to linguistic expressions, such as high and very low. Then, the weight of each risk factor was recalculated and ranked based on its importance by the Complex Proportional Assessment (CORPAS) risk assessment method. In [86], a fuzzy rule-based system was proposed for prediction of COVID-19 cases based on 11 input variables (e.g., cough, fever, and age) and 15 fuzzy rules. In [87], a hybrid-statistical advanced-fuzzy-cognitive-map algorithm was employed to predict COVID-19 cases based on data from Greece, South Korea, and Germany. The novelty of this FCM was the division of variable-concepts into three categories, namely, fuzzy concept states, inputs, and outputs. These concepts were defined by experts and include ten input concepts (such as restriction policies, number of patients per population size, intensive care units per population size, and others), two state concepts (impact of restrictions and healthcare system effectiveness) and one output concept (prediction). Then initial weight connections between concepts are proposed by the authors and then further optimized by fitting the model to the epidemic curve during a simulation period of 3 weeks.

In addition, epidemiological decision-making frameworks have been explored to combat waterborne diseases. Earlier methods utilized the exponential model and proposed risk factors of waterborne adenovirus transmission in potable water [88]. In [89], a stochastic RA model was introduced to evaluate the risk of infection through fecal contamination of recreational waters motivated by the outbreak of Cryptosporidiosis in Ontario. Three different types of recreational waters were considered, i.e., rivers, lakes, and swimming pools. The model considered input variables as input estimates or probability distributions derived from statistical analysis of literature-derived data. Due to lack of diverse data, the concentration of pathogen in swimming pools was estimated through a proposed mathematical model using the pathogen concentration following a fecal contamination incidence, the concentration of the pathogen in 1 g of liquid stool, and the volume of the pool. Similarly, the estimated exposure dose for Cryptosporidiosis through swimming was estimated based on another proposed mathematical model. In addition, the risk of infection was based on an exponential dose–response model but with the addition that it was based on multiple exposure events for each individual.

More recently, the effect of virus transmission from rivers to tap waters was explored by analyzing the water in three different river sites, inferring the risk factors, which were then incorporated in a disease transmission exponential model [90]. Moreover, pathogen transmission through wastewater and its effect on the communities has been studied in [91], where a water-based epidemiological framework was proposed that combines social and disease-related factors with chemical analysis of compounds in wastewater.

Knowledge-based approaches have also been considered to assess the risk related to small-scale disease spreading, such as in buildings. Airborne infectious diseases have been modeled using the Wells-Riley and dose–response models [92, 93]. In [94], the Wells-Riley model was coupled with a blockchain-based alert system for the RA of disease transmission in buildings. In addition, CO2 monitoring was proposed as a proxy for the COVID-19 transmission risk in [95].

During the early days of the COVID-19 pandemic, transmission-specific information was not clear even among experts. Therefore, certain studies attempted to estimate transmission and fatality factors by statistically analyzing epidemic data, such as those from the Diamond Princess cruise ship [96–99]. These methods could be helpful if incorporated in RA tools to predict epidemic outbreaks of recently discovered communicable diseases, especially in maritime scenarios. To the best of our knowledge, RA tools related to epidemic outbreaks in ships have been based only on knowledge-based methods [62–64]. An early study identified risk factors related to waterborne diseases associated with ships by analyzing data from previous epidemic outbreaks; however, these data were not incorporated in an RA tool [60]. In [62], an RA framework was proposed to assess the risk of epidemic outbreaks in ships based on the ports that they have visited. The risk of an epidemic outbreak was estimated based on a rule-based system that was employed to assign a level of risk to each ship, i.e., low, medium, and high risk. The system utilized a proposed cumulative exposure index (CEI) for each ship to assign a level of risk, by comparing the CDI to a reference index inferred from the Wuhan outbreak data (e.g., high risk when CEI is higher than the reference index). This index for each ship was provided by a mathematical model that calculated the infection exposure during the last 14 days based on the population density and the growth factor of the visited countries. The growth factor was a proposed mathematical equation that accounted for the infection risk in each country based on the daily change of confirmed cases.

More recently, an RA framework introduced the use of a probabilistic model considering four risk factors to calculate the infection risk for the entire ship, as well as for each ship compartment and space. The risk index considered a cumulative probability of a set of diseases to be present in the ship. For each disease, a multiplicative probability was calculated by considering the probability for the disease to be present in the ship and the cumulative probability of the disease to be present in each area of the ship. The probability of a disease to be present in a ship location was affected by the probability of the disease to result in an epidemic outbreak based on the location and transmission routes of the disease, as well as the effect of exposed passengers. The calculated risk index was then used to assign levels of risk based on a certain risk index threshold; in particular, three risk levels were defined, i.e., green (low), yellow (medium), and red (high) [63]. In addition, a conceptual classification of ship compartments in a cruise ship was proposed by the authors of the study. Similarly, in [64], an RA tool classified ship zones into three levels of infection risk (i.e., low, medium, and high). The RA tool utilized a knowledge-based rule system that was proposed by the author based on the relevant literature, and considered risk factors, such as poor ventilation and overcrowding. The system assigned risk factors to each area of the ship and classified them based on the number of risk indicators accumulated in each area.

4.2 Machine Learning-Based Methods

Moving beyond knowledge-based methods, ML provides state-of-the-art assessment tools without relying solely on expert knowledge that might be subject to personal biases and errors, or even without the need of manual feature extraction from relevant scientific evidence [100]. Epidemic-based RA ML methods have focused on large-scale scenarios utilizing spatiotemporal data, such as in [101]. In that study, a logistic regression model was used to create an RA tool for predicting risk of COVID-19 to occur in areas of the Guangzhou City, China. In detail, the logistic regression model was trained based on nine factors such as population density (permanent residents), population flow (population migration), COVID-19 infection distribution, and traffic density. This regression model was incorporated in an exponential model to calculate probability of infection occurrence in an area. Geographic detectors are statistical methods that are employed to infer spatial associations of impact factors in a geographic area. In this study, four geographic detectors (factor detector, interaction detector, risk detector, and ecological detector) were proposed to find spatial correlations between the proposed risk factors and the risk of disease transmission based on spatial epidemic data in each area of the city.

There has also been a plethora of spatiotemporal-dependent methodologies employing deep learning models for forecasting COVID-19 cases mainly focusing on LSTM networks [102], and fusion of spatial-aware models, or Convolutional Neural Networks (CNNs) with LSTMs [103, 104]. In [102], a multivariate LSTM was used to assess risk of COVID-19 spread by predicting the number of infected cases, deaths, and foot traffic in US counties. The input features considered in this study were the same as the output features. Another method of training deep learning methods was considered in [104], where Convolutional LSTM (ConvLSTM) was trained based on image frames representing overlapping areas of a geographical location. Each image frame included two channels; one channel represented the susceptible and the other the infected population at each pixel of the image frame. Furthermore, a cellular automaton (CA) model in [103] was combined with an LSTM network to predict COVID-19 cases. Specifically, the LSTM network inferred the transition probabilities used by the CA model to predict the disease spread. The CA model represents a geographic area, which in the context of [103] was applied for China, and each cell unit of this model considers a geographic location along with other features that include the population density, the migration index, and the number of COVID-19 incidents. Moreover, the transition probabilities utilized by the CA model were based on relative growth infectious rate, recovered rate, and death rate. These were calculated based on the confirmed, recovered, and death cases predicted by the LSTM network. The input features of the LSTM network were the same as the output features. A Bayesian ML-based model was combined with a spatiotemporal-related deep LSTM network to predict COVID-19 cases in the community of Castilla-Leon, Spain [105]. The LSTM network was used to predict the number of infections based on the previously observed infectious cases. The predicted cases were used in the Bayesian model as the expected value. In addition, the Bayesian model considered the correlation of disease cases and mobility data, the impact of population demographics (e.g., economic status), as well as the spatial correlation based on location of individuals.

Reinforcement learning (RL) has been proposed as an alternative method to predict the spread of COVID-19 [106, 107]. In [106], a Q-learning-based framework utilized a variation of the SEIR model that introduced three subcategories, namely, mildly infected and severely infectious, exposed, and recovered individuals. The SEIR model was used as the environment response to the control actions imposed by the RL-agent. The actions were divided into non-pharmaceutical actions, such as restriction on social gatherings, and pharmaceutical interventions, such as use of masks. The reward was calculated based on the reduction of infectious agents, the cost associated with implementation of the control actions, and the hospital saturation capacity. Similarly in [107], a RL-framework utilized a large-scale ABM to search for optimal control measures for mitigation of both disease spread and cost of intervention policies. The method considered epidemiological-related information, such as the numbers of infected cases and vaccinated individuals. Thus, a RL-framework can optimize the decision-making process for managing an epidemic outbreak, as well as assessing the risk of an epidemic outbreak.

4.3 Hybrid Methods

Methods combining ML algorithms with knowledge-based models can be categorized as hybrid methods. Hybrid methods can be divided into methods that refine a set of rules provided through the assistance of experts, historical data, or relevant literature review of the subject [108], and methods that combine RA tools with interpretable/explainable ML [109, 110]. This way methods benefit from the interpretability/explainability of knowledge-based systems and the power of ML approaches [111]. In [112], a hybrid method that fused data cases, population movement, meteorological, and socioeconomic data with spatial clustering analysis and geographically weighted regression was employed to identify the most correlated factors to COVID-19 spreading. In [113], a multiple linear regression analysis model based on experts' input was incorporated in a spatiotemporal decision support system. In [114], CNNs were used to identify West Nile River virus outbreaks and then the weights and biases of the model were used to infer five risk factors. Decision trees can also be utilized to infer the most influential risk factors. For instance, in [115], a regression tree was used to find the so-called actionable features with the largest variances related to the spread of COVID-19 in counties of the United States of America. Subsequently, these selected features were used as input to a LSTM-GRU (gated recurrent unit) hybrid network for the prediction of cumulative case and death trajectories. In [116], an autoregressive integrated moving average model was incorporated with a wavelet-based model to forecast COVID-19 cases and a regression tree was utilized to find critical parameters related to case fatality rate. Furthermore, disease spread among farm animals could also provide important information for probable epidemic outbreaks. For example, in [117], an interpretable ML pipeline, including SVM, random forest, and a GB model, was implemented for the RA of disease spreading in farms. In [118], an epidemiological diagnostics system that combined ML methods, fuzzy logic, and game theory was proposed. The use of Bayesian networks (BNs) has been recently explored [119], where an initial ML-based BN is generated from knowledge-based identified risk factors and optimized based on expert knowledge.

Hybrid methods have also utilized fuzzy logic by employing adaptive neuro-fuzzy inference systems (ANFIS) [120, 121] and neuro-fuzzy inferred rule-based systems [122]. In [121], a variation of SEIR model was used to predict the peak COVID-19 cases by utilizing a GA to calculate the probability of infection. In addition, an ANFIS was used for short-term forecasting of cases based on historical data from the past 10 days. It was implemented by using the first order-Sugeno model with one input (i.e., day number), one output (i.e., infected cases), and two fuzzy rules. Similarly in [120], a first-order Sugeno-based ANFIS was used to predict COVID-19 cases in India, by employing a Takagi-Sugeno type fuzzy rule system. The network used two input features, namely, COVID-19 cases and population demographics (e.g., size of population, population density, and median age). Moreover, in [122], a fuzzy rule-based system was implemented to predict daily cases of COVID-19 in Turkey. The system considered six input variables, such as the number of intensive care patients, intubated patients, and daily tests. The rules of the system were extracted based on the Wang-Mendel method. Table 2 summarizes the literature review of epidemic RA tools and methods presented in this section.

4.4 Risk Management

Another important field of study is the management of an epidemic outbreak and the mitigation of disease transmission [124]. Studies focused on epidemic outbreaks in the ship industry have highlighted the importance of effective control measures [1, 60, 125]. The most common type of diseases in maritime vessels are respiratory (airborne) and gastrointestinal (waterborne) infections [125]. The role of air circulation regarding airborne-based epidemic outbreaks have been greatly emphasized [93, 126, 127]; thus, proper ventilation and air cleaning devices, such as HVAC systems with high-efficiency particulate absorbing (HEPA) filters and ultraviolet germicidal (UVGI) devices, are considered as a major control measure of airborne diseases [25, 28, 95, 124, 128].

To combat waterborne diseases, control measures are focused on maintenance and surveillance of every step in the water distribution system, including potable, recreational, and waste disposal systems [25, 28, 60, 129]. Monitoring of potable water sources, i.e., trusted sources of water in ports [60], and even the monitoring of all water output sources [130] has been proposed to mitigate the risk of epidemic outbreaks. Interestingly, a study also underlined that ballast water could be a potential source of waterborne diseases [131]. Except for maintaining the quality of the water systems, retrofitting of the water distribution systems has been suggested in [124]. Proper water filtration and disinfection is another control measure for risk management of waterborne diseases [60]. In [130], the additional enzymatic treatment of water tanks has been proposed, and in the case of ferries, inspection of the systems by an external specialized team has been recommended.

In general, early detection of infected individuals is important to mitigate the transmission of diseases [132]. Regular testing and symptom screening has been suggested for crew members and passengers of ships during epidemic outbreaks [124, 133]. This can also be used as a preventive measure when it is applied in a pre-boarding scenario [61, 124, 134]. During an epidemic outbreak, effective control measures, or the so-called actuations, include the proper health education of individuals regarding signs and symptoms, as well as hygiene etiquettes (wash/disinfect hands, cough etiquette) and use of protective equipment such as masks and gloves [124, 128, 133]. Proper training, vaccination, health status registration of crew members, as well as establishment of emergency protocols for epidemic outbreak responses are some other preemptive measures [18, 25, 28, 125, 128]. On-ship medical centers have been proposed as an actuation during outbreaks [18, 132], and the incorporation of permanent medical facilities has been suggested [124].

Table 2 Overview of RA methods related to epidemic outbreaks. The references in bold refer to RA tools applied in ships

Studies	Models	Category	Parameters
[62, 63, 73–79, 88–90, 94, 95]	Mathematical models (SIR, SEIR, dose–response, Wells-Riley, gravity model, ST/DT)	Knowledge-based	Population demographics (age, economic status), healthcare system characteristics, intervention mechanics, spatiotemporal data, country-specific infrastructure capabilities, confirmed infected cases, pathogen concentration, population flow
[47, 64, 66, 67, 69–71, 91]	Rule-based systems	Knowledge-based	Experts' knowledge, data from previous outbreaks (rate of infection), infected and susceptible population, disease transmission mechanics (pathogen incubation period, transmission routes), healthcare system characteristics
[85]	IVq-ROFSs-CORPAS	Knowledge-based	Experts' knowledge, weighted risk factors (people, technology government, economy-related factors) with weights represented as linguistic expressions
[86]	Fuzzy rule-based system	Knowledge-based	Triangular membership function for each fuzzy variable (fever, cough, age, diabetes, travel history, breathing problem, flu, hearing problem, loss of smell, ache, and sore throat). Fuzzy rules combining two or more fuzzy variables
[123]	FCM	Knowledge-based	Linguistically expressed weights for fuzzy concepts, ten input concepts (restriction policies, number of patients per population size, intensive care units per population size, daily deaths, quality of healthcare system, medical staff, intubated patients, days since first death, rate of increase in deaths), two state concepts (impact of restrictions and healthcare system effectiveness), and one output concept (prediction)
[101]	Logistic regression	Machine learning	Number and location of confirmed infected cases, size and location of total population, population flow, population, hotel, shopping mall and public train station density, fever clinic, and supermarket distance
[102–105]	CNN, LSTM, ConvLSTM	Machine learning	Number of infections and deaths per day, susceptible population, geographic location of population, population density, migration index, mobility data, population demographics (e.g., economic status)
[106, 107]	Reinforcement learning, Q-learning, SEIR, ABM	Machine learning	Infected, exposed, recovered, susceptible population, location of population, transmission rate, intervention policies, hospital saturation capacity
[84, 112–118, 120–122]	Linear regression, CNN, decision trees, SVM, LSTMs, ANFIS, Neuro-fuzzy systems, BN	Hybrid	Experts' knowledge, number, and location of infected and susceptible population, population density, median age, population movement, wind speed, temperature, humidity, atmospheric pressure, precipitation, air quality, socioeconomic data (proportion of elderly people, number of hospital beds, health technicians and high education institutions, gross domestic product per capita), date of prediction, number of intensive care patients, intubated patients, daily tests

Quarantine of infected individuals has been a key actuation against the spread of communicable diseases. For example, quarantine was utilized during the early stages of COVID-19 in ship epidemic outbreaks, e.g., in the Diamond Princess cruise ship [135]. Studies suggest that isolation is a successful control measure [18, 61], whereas others argue that disembarkation is preferred to a quarantine policy [125, 128, 132]. In [133], passengers and positive cases were evacuated and some crew members remained in the ship, while being segregated in essential and non-essential personnel. Essential workers continued operations, while nonessential were quarantined in passenger cabins. Ship zoning according to the levels of contamination has been proposed in [25]. Similarly, a hybrid method of ship zoning and batch disembarkation policy has been suggested in [124].

Vessel design is important for mitigating the risk of epidemic outbreaks in ships. Some considerations include less and larger cabins, as well as more dining areas (usual sources of disease transmission) that allow social distancing [25, 28, 134]. Increasing the number of entrance/exit areas to avoid congestion of individuals has also been suggested [64]. Also, the use of telemedicine [124, 125], remote health management centers, smartphones, and the Internet-of-Medical-Things (IoMT) could provide helpful alternatives to manage epidemic outbreaks in ships [18, 36, 77].

5 Simulation of Disease Spread

To model the airborne, waterborne, and surface/fomite-mediated transmission of infections, a wide variety of methods has been employed. A review on the role of computational fluid dynamics (CFD) tools on the investigation of pathogen transmission can be found in [136]. These tools include the Finite Volume Method (FVM), Finite Difference Method (FDM), and Finite Element Method (FEM), which are commonly used commercial methods that can solve most fluid-related problems, as well as particle methods, including Smooth Particle Hydrodynamics (SPH), Moving Particle Semi-implicit (MPS) method, Finite Volume Particle (FVP) method, and the Lattice Boltzmann Method (LBM). Other methods that have been employed to model the transmission of waterborne pathogens through small droplets suspended in the air that can be inhaled include the Volume of Fluid (VOF) and discrete phase model (DPM) methods [137–139]. As regards the transmission through surfaces, the review on the relevant literature reveals that this is primarily modeled via mechanistic and/or stochastic models [140, 141].

While the above modeling approaches are used to simulate the local transmission of diseases, i.e., from a contamination source to people or among individuals, there is another category of models that are utilized to simulate the spread of disease at larger scales. According to a recent review study [142], there is a variety of simulation models that have been utilized during outbreaks to simulate disease spreading at a larger scale, capture relevant driving factors, predict risks and turning points, facilitate the optimal allocation of resources, and design and evaluate public health policies. These models include compartmental [74] and other mathematical models [78, 92], discrete event simulation (DES) [143], agent-based models (ABMs) [144], data-driven modeling approaches [91], and ML techniques [102]. The above models have been used to model infectious diseases including tuberculosis [145], rabies [50], porcine reproductive and respiratory syndrome virus [117, 146], influenza [147], West Nile River virus [114], Ebola [74], avian influenza [68], adenovirus [88, 90], *Escherichia coli* [90], Zika [75], SARS (severe acute respiratory syndrome) [148], MERS-CoV [149], and SARS-CoV-2 [73, 78, 79, 106].

Compartmental models are simplified mathematical constructs, which often use a system of ordinary differential equations (ODEs) to model and simulate infectious diseases. These models divide the population into labeled compartments, each represented by an ODE of the system. Each individual in the population can transit in-between, or progress through, these compartments during every timestep based on stochastic transition rates. These transition rates are established based on prior knowledge of previous outbreaks or inferred based on statistical analysis of the disease mechanics [150]. The most common compartmental model is the Susceptible-Infected-Recovered (SIR) model and its extensions. They are simple and easy to implement in different simulation scales, enabling faster response during an outbreak.

These mathematical models, despite their ability to accurately represent the mechanics of an infectious disease, fail to consider the spatial and social aspects that contribute to an epidemic outbreak. Agent-based modeling is the most widely used computational modeling approach that is employed to tackle this problem. More specifically, ABMs are spatial discrete-time simulation models, in which agents represent interacting actors or items of interest [144]. Their main advantage is that they can be combined with epidemiological mathematical models to simulate and visualize heterogeneous complex systems. Thus, they can be used to evaluate intervention policies and risk of epidemic outbreaks [107]. Statistical models are another category of mathematical models employed to simulate infectious diseases. In detail, these data-driven methodologies either consider statistical analysis of case-specific data to infer risk indices of epidemic outbreaks [79] or probabilistic models that estimate risk of transmission on large-scale (e.g., national level) [71] and small-scale (e.g., ships) [93] scenarios. Such data-driven methods can enhance the capabilities of the simulations by considering more complex dynamics specific to the population examined (e.g., healthcare system characteristics, meteorological data, population demographics) [75].

Some of the aforementioned simulation models have been combined with ML techniques to accelerate the process of identifying risk factors and decision making [113]. For example, an approach to estimate the infection risk was based on the estimated pathogen concentration in the vicinity of the individuals [151].

The rest of this section provides a review of the state-of-the-art models for the simulation of airborne, waterborne, and surface transmission of pathogens that could be considered in the context of a disease monitoring and control system, as described in Sect. 2.

5.1 Airborne Transmission Modeling

Airborne disease transmission refers to the spread of infectious agents, such as viruses, through the air, and is considered the primary transmission mode for several infectious diseases [151]. For example, the relative contribution rate of aerosol particles on the Diamond Princess cruise ship was assessed as the most dominant infection route for COVID-19 transmission [93]. More specifically, airborne transmission primarily occurs in confined spaces when exhaled respiratory droplets remain suspended in the air and can be potentially inhaled by other individuals [152]. Therefore, the risk for infectious disease outbreaks in indoor spaces is generally higher, since the concentration of contaminants is more likely to increase, considering the factors of insufficient ventilation and confined prolonged human contact [153, 154].

Notably, the quality of the HVAC system in an indoor setting is crucial for the spread of infectious diseases, since it regulates the movement and advection of any aerosols, pollutants, and carbon dioxide (CO_2) [155]. Furthermore, [156] reported that low humidity, low temperature, and a lack of air filtration constitute contributing factors to disease spread. Hence, measuring IAQ is crucial for determining whether the airflow of the HVAC system can mitigate the risk of airborne transmission [157]. This can be achieved by implementing different strategies such as air filtration, ventilation control, humidity and temperature regulation, as well as monitoring of environmental parameters [158, 159].

Although technical interventions can be employed to limit airborne transmission, modeling the dynamics of disease spread is essential for the development of effective prevention strategies [160, 161]. Modeling the evaporation and dispersion of droplets is a challenging and complex task when considering the multiphase interactions and variability of airflow conditions in an indoor setting. In this regard, CFD models incorporate these elements of variation to simulate respiratory actions, such as coughing and sneezing, to assess their effect on disease spread. There have been several studies that utilize CFD modeling to provide detailed information of the spatial distribution of infectious aerosol particles indoors.

For instance, a recent study investigated the transport, dispersion, and evaporation of saliva particles arising from a human cough [162]. More specifically, the proposed methodology incorporated a three-dimensional model based on fully coupled Eulerian–Lagrangian techniques that takes into account the relative humidity, turbulent dispersion forces, droplet phase-change, evaporation, and breakup parameters. The study showed that when someone coughs, the wind speed in an open space environment significantly affects the distance airborne droplets travel. Other studies focus on assessing the effect of ventilation and face masks under various airflow settings, to further enhance our understanding of droplet dynamics in these conditions. These studies, for instance, utilize CFD models to simulate aerosol plume dynamics [163] or examine the fluid dynamics phenomena that affect face mask efficiency under a coughing incident [164]. Furthermore, there are recent studies in the literature that focus on simulating long-range transport of infectious aerosols in ventilated indoor environments. One such example is explored by the authors in [165], where they investigated the impact of air ventilation systems and aerosols in an elevator under different airflow settings and found that the placement and design of the air purifier and ventilation systems significantly affect droplet dispersion. Additionally, other studies assess the possibility of airborne transmission in realistic indoor environments, for example, by measuring the ventilation rate in a restaurant using the tracer gas concentration decay method [166], or by investigating aerosol transport and surface deposition in a realistic classroom environment using computational fluid-particle dynamics simulations [167]. In a similar regard, there are studies that aim to identify the transmission mechanisms of respiratory aerosols and assess strategies for risk mitigation aboard an enclosed bus environment [168, 169]. These studies combine experimental, numerical analysis and CDFs to simulate particle evaporation and transport. However, despite the potential usefulness of CFD models, one major challenge is their high computational cost and their limited capacity to capture the full complexity of a given system. Another simulation method is Lagrangian Particle Tracking (LPT), which has been used to track aerosol hazards in operating rooms [170]. LPT methods can calculate particle trajectories and forces applied on them, such as gravity, buoyancy, drag, and inertia that can alter their velocity and direction in the flow. LPT can adequately simulate the motion of a discrete number of particles, but it is considered computationally expensive.

An alternative direction to monitor disease transmission through the airborne route is to employ ABMs to simulate the movement and behavior of individuals within a predefined computational environment [171]. Several studies have employed ABMs to investigate disease transmission and exposure risk in various indoor settings, such as airplane cabins [172], fever clinics [173], subway stations [172], and restaurants [174]. Moreover, certain studies have specifically focused on developing ABMs for aerosol-based airborne transmission [175, 176], while others address disease forecasting [177]. Notably, ABMs are considered computationally intensive and challenging to calibrate as they rely heavily on prior assumptions.

CFD models can accurately approximate real-life scenarios of airborne transmission using equations related to airflow based on experimental studies [151]. Nevertheless, simulations of airborne diseases that can be transmitted through human-to-human interaction need to consider the human factor in order to assess the risk of epidemic outbreaks in social scenarios [178]. This results in scenarios where the CFD models are required to recalculate the airflow field to account for exposed or infected individuals and the spatial changes of the simulated space through time, which are affected by people movement. The need for recalculations leads to computationally expensive simulations that are impractical [179]. Therefore, in order to mitigate this effect, researchers have proposed hybrid methods, which incorporate probabilistic models and ABMs.

Probabilistic-CFD approaches are usually employed to find the probability of infection of individuals based on dose–exposure models [146, 180, 181], the Wells-Riley model [145, 147, 148, 179, 182], and other probabilistic models such as the Markov Chain Model (MCM) [183] or probability distribution functions (PDFs) [184]. The effectiveness of such models has been presented in studies that compare CFD-based and probabilistic-CFD methods [182, 185]. Probabilistic-CFD models consist of two parts, i.e., the exposure equation that calculates the exposure dose of individuals based on the concentration of pathogens provided by CFD simulations and the infection probability equation, usually an exponential model, which estimates the probability of infection based on the calculated dose exposure and other infection-related parameters [186]. These infection-related parameters can either be determined based on previous research, e.g., experimental or CFD studies [147], or by statistical analysis, e.g., maximum likelihood estimation or MCM [145, 146, 187]. Probabilistic models such as the Wells-Riley model assume that the pathogen concentration is uniformly distributed, which is less accurate in realistic scenarios [188]. In [184], the use of a PDF was proposed as a solution to this problem. More specifically, the parameters of the PDF, e.g., mean and standard deviation of pathogen concentration, were determined by statistically analyzing 174 high-resolution CFD simulations. Such probabilistic methodologies mostly focus on predicting infected individuals without considering social factors and spatial changes of the environment. A solution to this problem has been proposed in [179], where a spatial flow impact factor was incorporated in the Wells-Riley model to account for spatial changes. Furthermore, this approach could be used for minimizing the overall infection rate at a specific location by optimizing the location distribution of population and facilities.

ABM-CFD approaches introduce motion mechanisms, thus providing more realistic simulations of airborne disease transmission scenarios [144, 188]. These methods utilize CFD simulations to predict the airflow field, such as droplet distribution and then incorporate the inferred pathogen concentration to the ABM that predicts infected and exposed human agents based on deterministic, probabilistic, or epidemiologic dynamics [144, 173, 188–190]. Approaches considering a deterministic disease transmission model consider an infection occurrence based on thresholds determined by experimental studies or relevant literature [174, 188]. Non-deterministic approaches either utilize probabilistic infection models, such as the dose–exposure model [173], or epidemiological models, such as the Susceptible-Infected-Removed and Susceptible-Exposed-Infected models [143, 190]. Based on the newly infected population, the ABM model updates the infection status and location of the population, which are then used as boundary conditions for the next time step of the CFD simulation [143, 173, 188, 189]. ABM methods that incorporate social features that occur in real-life scenarios, such as contact avoidance of agents, employ the Social Force Model (SFM) [144], which has been reported to provide high-fidelity simulation results [191]. Similar methods utilize computational crowd dynamics (CCD) models [174], discrete event simulation (DES) models [143], and clustering methods [189] to simulate spatiotemporal changes in the system. Moreover, some methodologies incorporate more sophisticated mechanisms in the ABM simulation, such as coughing, use of protective equipment, hand hygiene, vaccination, and different ventilation techniques [143]. In addition, several studies combine airflow parameters, e.g., airflow velocity, and agent-related variables, e.g., height, face orientation, and inhalation/exhalation cycle [188, 190]. In [190], the impact of agent movement in the airflow field of the simulation is considered by incorporating a wake model, namely, the Actuator Line Model (ALM), in the simulation.

ABMs can be simulated on 2D or 3D mesh grids. ABMs based on a 3D grid system can provide more fine-grained simulations that can result in more realistic results [190, 191]; nevertheless, they can become computationally expensive [188]. In contrast, an ABM simulation based on a 2D grid mesh could be more computationally efficient but may compromise the fidelity of the simulation due to interpolation of data from a 3D to 2D mesh grid system. In [174], a hybrid method was adopted, which utilized a CCD model based on a 2D-background mesh grid and a CFD model based on a 3D mesh grid to simulate pedestrian motion. The translation from 3D to 2D and vice versa was achieved by utilizing interpolation coefficients that were computed at the start of the coupled simulation.

Hybrid methods often utilize an Eulerian-based CFD or a Lattice-Boltzmann (LB)-based model to calculate the airflow and other analytical methods to find the trajectory and dispersion of droplets [178]. More recently, an LB-based method using a Large Eddy Simulation (LES) model was combined with an SFM-based ABM to simulate pathogen dispersion in evacuation scenarios [191]. In [188], a semi-analytical approach utilized a set of coupled ordinary differential equations (ODEs) to simulate the dynamics of droplets as a function of the airflow. In additional, mechanics involving exposure to pathogens take into consideration short- and long-range airborne routes [187], as well as direct inhalation of coarse droplets and direct deposition of medium droplets on facial membranes [173]. Another parameter that is considered in hybrid methods is the size range of droplets, since it affects the possibility of infection. Studies often define an interval-based size range [146, 180, 189] or a discrete set of size ranges [145, 181, 187], while others focus on one type of droplet size [148]. In addition, methods that study droplet deposition on surfaces could incorporate surface disease transmission mechanics to provide a more realistic disease transmission model [187]. A hybrid method could also incorporate variations of coupled long-range, close-contact, and surface dose exposure mechanisms in a dose–response model to predict risk of infection and disease transmission. Long-range airborne transmission could be calculated based on CFD simulations or the multi-zone infiltration and exfiltration model (MIX) and fomite transmission based on discrete-time non-homogeneous MCM. Furthermore, dose exposure related to close-contact interaction depends on short-range airborne transmission, direct deposition of large droplets on mucous membranes, and inhalation of droplets based on the normal distribution model [149].

In summary, the methods used to model airborne disease transmission comprise numerical physical models, mechanistic models, ABMs, or any combination of them. The novelty regarding the methods mostly focuses on the case study (e.g., COVID-19 transmission in buildings) and the parameters regarded in the models (e.g., ventilation incorporated, specific droplet size, use of motion mechanics in ABM, different SFM parameters considered). Table 3 summarizes the above research, providing a comprehensive overview of the literature regarding airborne transmission modeling.

Table 3 Overview of key studies reported in the literature regarding airborne transmission modeling

Study	Methods	Novelty	Key parameters	Case study	Validation
[162]	CFD, Eulerian –Lagrangian	Assessed transport, dispersion, and evaporation of droplets after cough	Relative humidity, turbulent dispersion forces, droplet phase-change	Computational mesh/Airborne diseases	–
[163]	CFD, FANS, RANS & Infectivity models	Coupled CFD pathogen concentrations in expelled aerosol plumes with infectivity models	Exposure duration, spatial-temporal aerosol concentrations, viral load	Simulated mesh domain/Airborne diseases	✓
[164]	CFD, Eulerian –Lagrangian	Proposed criterion for calculating mask efficiency, considering penetration dynamics	Turbulent dispersion forces, droplet–droplet and droplet–air interactions	Computational mesh/ Airborne diseases	✓
[192]	CFD, Krylov-Eulerian model	Assessed long-range transport of aerosol particles and reduce the computational cost with ROM	Mean velocity and coefficients of turbulent transport	Classroom/Covid-19, influenza, rhinoviruses	✓
[165]	CFD, Eulerian–Lagrangian	Assessed air purifier on droplet dispersion under three flow scenarios	Placement and design of air-purifier, evaporation, droplet–droplet interactions	Elevator/COVID-19	–
[193]	CFD, Wells-Rilley	Assessed likelihood of secondary infection without measuring outdoor air supply rate	CO_2 concentration	Building/Airborne particles	✓
[171]	ABM, MC-LTTA	Computed daily probability of close contact between agents	Individual schedules, occupancy level, pedestrian traffic	Building/COVID-19	–
[172, 174]	CFD, FEFLO, PEDFLOW & CCD	Coupled CFD and CCD for high-fidelity pathogen transmission and infection simulations.	Temperature, smoke, toxic substances	Airplane cabins, subway train stations, restaurant/Aerosol pathogens	–
[173]	ABM & Infection transmission model	Studied effects of patient influx and layout of fever clinics on cross-infection risk	Activity trajectories, inject flow, movement speed	Fever clinic/COVID-19	✓
[194]	ABM, GCVM	Simulated pedestrian dynamics with microscopic model and defined an index for customer contact	Degree of contract, position, moving direction and speed	Supermarket/COVID-19	–

[93]	DRM	Assessed aerosol particles as the most dominant infection route for COVID-19 transmission	Incubation and subclinical infectious period, aerosol ratio	Cruise ship/COVID-19	–
[153]	CFD, Wells-Riley	Estimated an average quanta emission rate of infectious aerosols from a single index case	Volumetric breathing rate, loss rate due to ventilation and deposition onto surfaces	Church hall/COVID-19	–
[195]	CFD, Wells	Determined the airborne transmission of COVID-19 through droplets in suspension	Droplet size, water partial vapor pressure, air density, air kinematic viscosity	Classroom/Airborne diseases	✓
[167]	CFD, RANS	Assessed aerosol removal and surface deposition	Particle size, aerosol source location, glass barriers/windows	Classroom/Aerosols	–
[156]	Mathematical & Multivariate analysis (PCA, SOMs)	Simulated the virus load released by an individual, with a range of microdroplet emissions	Emission and respiratory rate, viral load	Hospital ward/COVID-19	–
[166]	CFD, k-ε turbulence model, tracer gas model	Simulated the spread of fine exhaled droplets, by the index case and airflow pattern	Temporal droplet concentration, ventilation rate	Restaurant/COVID-19	–
[169]	CFD, RANS	Combined experimental and numerical analysis and adopted a risk metric	Flow field, size distribution, and aerosol response time	Bus/COVID-19	–
[168]	CFD, RANS, LES	Assessed evaporation and transport of solid-liquid mixed droplets with thermal manikins	Humidity, droplet size, dispersion	Bus/Respiratory diseases	✓
[181]	Multiphase Eulerian frame. Transient Eulerian species transport equation. Dose–response model	Incorporated effect of ECS in the simulations	Expiratory aerosol particles dispersion and deposition, effect of ECS	Aircraft cabin/ Influenza	✓
[146]	CFD using ANSYS Fluent, stochastic dose–response model.	Examination of three stochastic dose–response models	Dose of infectious pathogen, number of infected and exposed pigs, particle size, airflow, and particle density	Animals in confined spaces/PRRS	✓

(continued)

Table 3 (continued)

Study	Methods	Novelty	Key parameters	Case study	Validation
[180]	Eulerian-based CFD, LPT, exponential dose–response model	Examined the impact of different ventilation strategies	Properties of droplets such as size, velocity, position for the CFD simulation	Office rooms/Airborne diseases	✓
[148]	CFD using ANSYS Fluent, drift-flux, particle model, spatial-aware variation of Wells-Riley	Spatial-aware variation of Well-Riley and integrated in CFD simulations. Drift-flux particle model to calculate exhaled pathogen concentration	Droplet size, exposure time, no death rate, evaporation of droplets	Hospital ward/ SARS	✓
[147]	CFD using ANSYS Fluent, deterministic and Wells-Riley	Simulated droplet dispersion based on single-cough and multiple-cough exhalations	Airflow, droplet dispersion, temperature, effect of masks	Aircraft cabin/ Influenza	✓
[145]	CFD using ANSYS Fluent, mix-ratio (Wells-Riley variation)	Proposed a mixed-ratio method for calculating particle count in breathing zone	Range of droplet size, ventilation, exposure time, cough frequency, passengers included in the CFD model	Aircraft cabin/ TB	✓
[179]	Eulerian based, LPT, Wells-Riley model variation with SFIF	Incorporated spatial flow impact factor (SFIF) in Wells-Riley model and combined with CFD	One infected and susceptible in every room, pulmonary ventilation rate, quantum generation rate and exposure duration	Hospital ward/ Airborne diseases	✓
[182]	URANS-based CFD, Wells-Riley	Case study, use of masks, and HVAC	Droplet size, HVAC system impact, breathing rate, of mask usage	Bus/Airborne diseases	✓
[183]	ANSYS Fluent, LPT combined with MCM	Calculated transition probability matrix based on CFD. Used Markov chain to calculate concentration distribution of particles	Airflow field, inertial effect of particles, droplet size	Clean room, Office room, first-class airplane cabin/Airborne disease	✓
[184]	RANS-based CFD, exponential dose–response model	Estimated probability of aerosol concentration based on a probability distribution (PDF)	Droplet size, ventilation rate, exposure time, room configurations	Indoor spaces/ Airborne diseases	✓
[187]	Steady-state CFD, MMBE (macroscopic mass-balance equation), variation of dose–response model	Used MMBE and discrete-time non-homogeneous MCM for surface contamination. Average PI is based on a least-squares fitting method	Concentration of droplets, deposition efficiency of droplets, viral load in exhaled droplets, filtration efficiency of mask, mask usage, droplet size	Bus/ COVID-19	–

[188]	LPT, Flow physics model, coupled ODEs, deterministic RI, ABM	Proposed a mesoscale ABM framework using SFM. Used semi-analytical method based on a set of coupled ODEs for dynamics of droplets	Airflow-related (e.g., airflow velocity constant flow), human agent variables (e.g., height)	Indoor spaces/ Airborne diseases	✓
[189]	LBM, LPT, ABM	Parallelized CFD simulation using the GPU. Clustering scheme for ABM model	Interactions of 1–5 individuals within 2 meters, velocity profiles droplet size, no particle-particle interaction	Supermarket/Airborne diseases	–
[190]	LES, RANS-based, SEI, ABM	Combined CDF with SEI-based ABM. ABM model affected CDF simulation every time step and vice versa	Exhalation-inhalation breathing cycle, face orientation, 3D location, breathing-related properties for every agent	Room-Level/Airborne diseases	–
[174]	LES, Runge-Kutta, second order Crank-Nicholson, deterministic RI, ABM	Combined CDF with a CCD model (PEDFLOW). CCD model calculated motion-related boundary conditions used in the next step of the CFD simulation	Social factor parameters such as, will force and contact avoidance, position/velocity of pedestrian, flow-field-related parameters	Restaurant, Train station/Airborne diseases	–
[143]	CFD based on BIM, SIR, DES, ABM	DES combined with SIR for ABM	Ventilation rate, temperature, exhaled breathing rate, volume flow rate, face covering, hand hygiene, vaccination	Building/COVID-19	–
[191]	LBM, LES, Scalar transport, ABM	SFM combined with a wake model (ALM) used for the ABM	Velocity force, repulsive force, body contact force, pollutant concentration, thermal diffusivity, pollutant diffusivity, fluid density	Meeting room, cinema, concert hall/Airborne disease	–
[149]	CFD, MIX, MCM, normal distribution model, ABM, dose–response	Incorporation of long-range, close-contact and fomite-mediated transmission. Different agent behavior models	Ventilation rate, movement and contact of agents, fixed behavior rules for each agent, droplet size, fixed viral load on droplets	Hospital/MERS-CoV	–

5.2 Waterborne Transmission Modeling

Waterborne transmission refers to the acquisition of disease by exposure to pathogens through contaminated drinking water, recreational water (e.g., pools), bathing, washing, or eating food exposed to contaminated water, and overall lack of clean water supply, sanitation, and hygiene. Water can become contaminated at the source, during transport to the storage facility, in storage containers, or via improper handling. Several studies have attempted to model waterborne transmission, and it has been shown that, apart from the direct introduction of pathogens into the body, i.e., drinking water and eating contaminated food, waterborne transmission takes place through aerosols containing contaminated water particles of different sizes.

It has been reported that stack aerosols are generated within vertical building drainage stacks during the discharge of wastewater containing feces and exhaled mucus from toilets and washbasins. Experimental measurements regarding the stack air pressure and temperature distributions indicated that stack aerosols can spread to indoors through pipe leaks, providing direct evidence for the long-range aerosol transmission of COVID-19 through drainage pipes via the chimney effect [196]. Another study reviewed the transmission of COVID-19 through aerosolized wastewater and aerosols generated by toilet flushing [197]. In [137], the VOF method was employed to simulate different flushing processes and the VOF–DPM method was used to model the trajectories of aerosol particles during flushing. The simulation results indicated the massive upward transport of virus particles, with 40–60% of particles being transported above the toilet seat, leading to large-scale virus spread. The VOF-DPM was also employed to investigate the transmission of infectious diseases in restrooms during male urination. The results revealed that male urination can induce strong turbulent flow with an average urine impinging velocity of 2.3 m/s, which can agitate virus particles and raise them [139]. A similar study was conducted regarding virus transmission from urinals. It was reported that the trajectory of the particles generated by urinal flushing exhibited an outward spreading mode, reaching a height of 0.84 m within 5.5 s [138]. In [198], CFD and DPM were used to quantitatively and visually demonstrate the effects of ventilation airflow speed, fan locations, and fan size on virus aerosol distribution due to toilet flushing in a typical family bathroom. They reported that the traditional ceiling fan was barely functional since aerosol particles were not adequately removed, while the side-wall fan functioned more efficiently, and its ventilation capability and performance were significantly better, removing nearly 80.9% of the lifted aerosol particles. Putting the toilet lid down before flushing is an effective method to prevent virus transmission. Nonetheless, male urinals do not have a lid or other barriers and people do not always have the habit of putting the toilet lid down before flushing to prevent aerosol cloud generation. In [199], both direct and indirect transmission pathways were incorporated into a reaction–diffusion waterborne pathogen model, considering the spatial heterogeneity and the mobility of host population and flowing water on a spatial continuous and bounded domain.

Mathematical modeling has also been utilized to estimate the risk of infection based on simulations of water-mediated pathogen transportation in water distribution systems. Waterborne disease transmission scenarios have been explored, calculating the probability of infection based on dose–response models or deterministic infection mechanisms [200, 201]. In [202], a spatially explicit network model was utilized to simulate short-term pathogen transport throughout communities. The network model contains nodes that represent human communities that are arranged in a spatial setting and are connected through hydraulic pathways. The evolution of pathogen transportation is monitored by an ODE epidemiological system that is affected by the population size (susceptible and infected individuals), pathogen concentration in the water supply, and human mobility. In [203], the impact of water and sanitation services on waterborne disease transmission was investigated. A variation of the SIR model was introduced to simulate a cholera outbreak in Haiti. The so-called susceptible–infected–water–dumpsite–recovered (*SIWDR*) model incorporated the impact of open dumpsites as pathogen concentration sites and deficits of water sanitation services in the SIR ODE system. The affected water systems included were the drinking water supply system and the wastewater and sewage system. For the dumpsite-related parameter, deficits in municipal solid waste management, and other non-water-related pathogen concentration sites were also considered in the model. In this model, the infectious population affected the number of concentrated pathogens in the dumpsite-related ODE, as well as in the water-related ODE. The dumpsite-related ODE also affected the water-related one, and both subsequently affected the number of newly infected population. Table 4 summarizes the above research, providing a comprehensive overview of the literature regarding waterborne transmission modeling.

5.3 *Surface Transmission Modeling*

Surface transmission, also known as fomite-mediated transmission, occurs when susceptible individuals come into contact with surfaces that have been contaminated by infected individuals through direct touching, direct deposition of respiratory droplets, and/or deposition of respiratory aerosols under different conditions [93]. Numerical simulations have been conducted regarding the virus deposition on surfaces via aerosol/droplet transport due to inhalation or coughing under different scenarios [204, 205]. However, the transmission modeling from contaminated surfaces to humans has been mainly focused on mechanistic or stochastic models. In [140], the contribution of fomite-mediated exposure to infection and illness risks during outbreaks was estimated. In particular, a stochastic simulation model in discrete time was developed to predict 17 h of simulated human behavior, taking into consideration hand-to-porous surfaces, hand-to-nonporous surfaces, hand-to-mouth, -eyes, -nose, as well as hand washing events. In another study [12], two mechanistic models of indirect transmission within the quantitative microbial RA (QMRA) framework were adopted to estimate the infection risk for COVID-19 in

Table 4 Overview of key studies reported in the literature regarding waterborne transmission modeling

Study	Methods	Novelty	Key parameters	Case study	Validation
[137]	VOF combined with DPM	Effect of toilet flushing on virus spread	Aerosol particle distribution, flushing pressure, vorticity	Vertical buildings/COVID-19	–
[138]	VOF combined with DPM	Effect of urinal flushing on virus spread	Discrete particle distribution, fluid velocity and vorticity	Urinals/COVID-19	–
[139]	VOF combined with DPM	Effect of male urination on virus spread	Discrete particle distribution, fluid velocity and vorticity	Toilets/COVID-19	✓
[198]	CFD combined with DPM	Effects of ventilation airflow speed, fan locations, and fan size on the distribution of virus aerosols in a typical family bathroom	Aerosol particle distribution, flow velocity, pressure, and vorticity	Typical family bathroom/COVID-19	✓
[199]	Reaction–diffusion waterborne pathogen model	Incorporation of both direct and indirect transmission pathways in a waterborne disease model	Spatial heterogeneity and mobility of host population, and flowing water characteristics	Direct and indirect waterborne pathogen transmission	–
[200]	Dose–response model	Optimal water heater temperature for reducing both costs and health risks	Water temperature, plumbing system parameters, infection and energy usage costs, scalding per showering event	Hospital hot water systems/Legionnaires' disease	–
[202]	Network model, SIR variation	Incorporated effect of human mobility and contaminated water supply in SIR	Population size, human mobility, pathogen concentration in water	Cholera	–
[203]	SIWDR	Incorporated effect of open-dumpsites and wastewater system defects in SIR	Open-dumpsite and water distribution	Cholera	✓

community settings and provide guidance on potential intervention strategies. More specifically, the first model was used to estimate the infection risk for single contacts with contaminated surfaces; the SARS-CoV-2 RNA concentration on surfaces was obtained from the literature and concerned surface contamination in public spaces, e.g., bus stations, gas stations, stores, and playgrounds. The second model was used to estimate risks from surface-mediated community transmission as a function of the prevalence of COVID-19 cases in the community and assess the efficacy of feasible intervention strategies of hand and surface disinfection. In [141], the transmission of pathogens from surfaces to fingertips was examined as an alternative way to simulate fomite-mediated transmission. The potential magnitude of exposure from surfaces to fingertips was calculated based on a mechanistic model. More specifically, the model calculated the pathogen concentration on the fingertips after contact with a contaminated surface based on the pathogen concentration of the surface before the contact, the pathogen concentration of the fingertips after the contact, and the transfer efficiency of the pathogen. The concentration-related parameters were calculated based on experimental data and statistically adjusted to account for the swabbing efficiency in the experimental study. In [206], comparative analyses on the routes of transmission of influenza A (H1N1), SARS CoV, and norovirus in an air cabin were conducted. A review on the mathematical modeling of fomite-mediated transmission, factors that affect transmission of microbes between fomites and humans, and the implications for human health can be found in [207].

Fomite-mediated transmission has also been simulated using mathematical models such as the Environmental Infection Transmission System (EITS) framework. The EITS is a compartmental model that incorporates live pathogens of the environment as a parameter in the SIR model. It was proposed in [208], where the epidemiological dynamics of the EITS are described by an ODE system that is affected from the traditional transition rates of infection and recovery, as well as environmental-related probabilities such as fraction of pathogens picked by a susceptible person, probability of infection based on pathogens picked, number of pathogens deposited to environment per unit of time, and rate at which pathogens are eliminated from the environment. The dynamics of the model can be simulated in a deterministic or a stochastic way, where events (e.g., a susceptible becomes infected) happen simultaneously and continuously or a single event randomly occurs in each case respectively.

An extension of the EITS was proposed in [209], where a model combined hand contamination due to fomite touching and infection based on self-inoculation, i.e., infection due to contaminated hands touching the mouth or other membranes. Table 5 summarizes the above research, providing a comprehensive overview of the literature regarding surface/fomite-mediated transmission modeling.

The numerical physical and epidemiological models reviewed in this section can become computationally expensive and introduce a trade-off between accuracy and efficiency, as their complexity increases. The following section introduces the notion of reduced-order modeling (ROM) that is often utilized to mitigate this problem, and describes relative AI-based methods that can be incorporated in simulations of disease transmission models.

Table 5 Overview of key studies reported in the literature regarding surface/fomite-mediated transmission modeling

Study	Methods	Novelty	Key parameters	Case study	Validation
[12]	Two mechanistic models of indirect transmission in the QMRA framework	Assessment of hand and surface disinfection effectiveness	Dose–response relationship, ratio of genome copies to infective virus	COVID-19	✓
[140]	Stochastic simulation model in discrete time and two dose–response models	Contribution of fomite-mediated exposure to infection and illness risk in outbreaks	Hand-to-mouth contacts, number of hand washes, hand-to-mouth transfer efficiency, fractional surface area (FSA) of hand-to-nose contact, FSA of hand-to-surface contact	Norovirus	–
[141]	Approximate Bayesian Computation combined with the QMRA framework	Impact of transfer efficiency modeling assumptions on estimated exposure	Viral concentration on fingertip, surface area of contact, swabbing efficiency, transfer efficiency	Rotavirus	✓
[206]	Multiroute disease transmission model/Mathematical model	Simulation of outbreak of three different in-flight infections in a similar cabin environment	Dose–response rate, shedding parameters, virus inactivation, touching behavior, transfer efficiency	Air cabin/Influenza A H1N1, SARS CoV, and norovirus	–
[208]	EITS	Proposed the EITS framework with deterministic and stochastic variations	Live pathogens in environment, pathogen decrease rate, deposit rate, pick up rate	Influenza	–
[209]	Variation of EITS	Extended the EITS model	Inoculation, fomite touching, excretion, pathogen inactivation, decontamination	Norovirus, influenza, rhinovirus	–

6 AI-Based Reduced-Order Modeling

Numerical simulations of high-order models and complex simulation scenarios can become computationally demanding as the complexity of the model increases. Reduced-order modeling (ROM) is a popular method that can mitigate this problem by reducing the dimensionality of the simulated model [210]. Traditional methods use proper orthogonal decomposition (POD) and principal component analysis (PCA) but are limited by their linear nature. Consequently, the need for nonlinear alternatives has resulted in the emergence of AI-based ROMs employing deep learning methods [211]. AI-based ROMs can be combined with other ML methods to generate surrogate models, alleviating the need for using numerical solvers [212]. Although AI-based ROM methods have been explored in various scenarios, including the modeling of airflow [211, 213], to the best of our knowledge they have not been yet explored in the context of disease transmission in ships.

More specifically, numerical simulations are affected by the density of the mesh that discretizes the flow domain and can become computationally expensive for high-order models [211]. ROM has been introduced to mitigate this effect by projecting the high-order model to a lower dimensional space. Traditional ROMs have employed projection-based methods such as POD to derive a reduced basis function that can be used as a proxy for the high- or full-order model [192]. These projection-based methods are considered linear dimensionality reduction techniques. AI-based methods have become popular due to their capability to model nonlinear characteristics. Despite their popularity, there is no evidence of any application in the domain of epidemic outbreaks in ships.

Autoencoders (AEs) are a family of neural networks that are used to map high-order features of models to accurate latent representations and vice versa [212]. AEs comprise an encoder and a decoder part, where the encoder part maps high-dimensional space features to latent representations and the decoder one is responsible to transform the latent representation to its original form [214]. Early methods utilized multilayered perceptron AEs (MLP-AEs) that incorporate nonlinear activation functions after each fully connected layer to reduce the dimensionality of high-order models [210, 215]. MLP-AEs lack spatial awareness and can become computationally expensive as the scale of the high-order model increases. Convolutional AEs (CAEs) have been proposed as a computationally efficient alternative to this problem [212]. CAEs include an encoder part with convolutional layers that extract spatial features from the high-order models [211]. Then, the extracted features are parsed through a fully connected layer to produce a latent representation as in MLP-AEs and the decoder part utilizes transpose convolution or up-sampling to infer the high-order model from the latent representation [216, 217]. In addition, adversarial AEs (AAEs) can be used to generate more accurate latent representations of the full-order model [210]. Physics-informed neural networks (PINNs) can be effective in simulation scenarios where the label data is limited and focused on a specific PDE/ODE [210]. For instance, in [218], a PINN framework was used for the reduced-order modeling of nonlinear dynamics.

Certain studies have proposed the use of AEs to reduce the order of a simulation model and employ reduced-based order solvers to calculate the results [219, 220]. Others try to completely bypass the solvers by training an AI-based model to predict the result of the numerical simulation acting as a surrogate model [213]. This can be either achieved by incorporating the numerical solver solution in the loss function [211, 213] or by using other optimization algorithms such as genetic algorithms (GAs) [214]. Numerical simulations, e.g., CFD simulations, consider temporal features to predict the output of the next time step. Long-Short-Term-Memory (LSTM) neural networks are neural networks that have been used to simulate the solution of a numerical solver, as they can retain temporal information for previous timesteps [210]. They have been proved successful in predicting accurate solutions for CFD simulations [213, 221].

ROM methodologies that incorporate the solution to the network utilize AE-based networks to generate a latent representation of the full-order model and then use deep learning methods, such as LSTMs [211]. Moreover, adversarial trained LSTMs have been utilized to predict solutions of a CFD simulation [222]. Studies that focus on fast simulations that depend on linear characteristics have explored the implementation of a CAE-extreme gradient boost (XGBoost) model [217]. Others that explored ROMs of both linear and nonlinear characteristics used an AE-radial basis function model [214]. Also, AEs that utilize LSTMs can become computationally expensive, thus spatiotemporal CAEs could be considered in such scenarios [212].

Epidemiological models have been incorporated in hybrid simulations combining ABM and numerical methods. There could be complex scenarios where the ABM part of the simulation could become computationally expensive. Thus, AI-based ROMs can be applied for dimensionality reduction of the epidemiological models, e.g., SIR. For instance, in [223], a PCA-based ROM was used to predict epidemiological dynamics of an idealized town by exploring the use of a bidirectional LSTM and a generative adversarial network (GAN) for the prediction part. More specifically, in the context of airborne transmission of respiratory infections, POD-based ROM approaches have been applied in indoor environments [192]. However, there has been a recent surge of methodologies that leverage neural networks as function approximators to model the dynamics of infectious diseases [224]. These approaches expand the application of SIR compartmental models by PINNs to forecast disease progression. For example, in [225], PINNs were employed to mathematically model and simulate the coupled dynamics of flow-mediated spreading of disease in enclosed spaces, such as buildings and aircraft cabins. Similarly, in [226], PINNS were utilized to optimize the estimation of the governing equation coefficients of a COVID-19 mathematical model, whereas in [227], their robustness and efficiency was demonstrated for 11 highly infectious diseases with various levels of complexity.

7 Research Challenges

The above review of methodologies aimed at exploring the aspects of a closed-loop system for disease monitoring and control in cruise ships, covering issues related to sensor technologies, transmission modeling, and risk assessment and management. This work has shed light on research challenges that should be taken into account in the development of systems for managing epidemic outbreaks on cruise ships. These challenges can be summarized as follows:

There is a lack of models that simultaneously incorporate information regarding the transmission of diseases via surfaces, air, and water to improve predictions.

Hybrid models that combine numerical models with ML/deep learning methods have not yet been explored in the context of disease transmission in passenger ships.

Fusion of all the complex dynamics introduced in Sect. 5.1 (e.g., SFM, human mobility, protective equipment and ventilation strategies) could provide insights on creating as accurate simulations as possible.

Simulations of disease transmission in ships could benefit from combining local (i.e., room level) and global (i.e., ship level) disease transmission mechanics.

The effect of all possible intervention policies mentioned in Sect. 4.4 has not been examined in methods simulating epidemic outbreaks in ships.

AI-based ROM can be utilized to provide computationally efficient simulations of disease transmission in ships.

Some infectious diseases, such as COVID-19, may spread silently among carriers without presenting symptoms. The automatic detection of possibly contaminated asymptomatic passengers is still an open problem.

Factors related to disembarkation at the different places (contact with non-passengers, crowded means of transport, e.g., shuttle bus, boats) have not been yet sufficiently researched.

There is a lack of a unified automated control system that combines multidisciplinary research for the continuous monitoring and assessment of IAQ.

There is a need to inform model predictions with explanations regarding the underlying factors leading to decision making.

There is a need for methods providing insights regarding the uncertainty of predictions of automated disease spread control systems under dynamic environments.

8 Conclusions

This chapter investigated the topic of disease spread control in cruise ships with computational methods, and it introduced the concept of a *closed-loop* model for this purpose. This is a generic model with a modular architecture composed of sensors to monitor ship areas, a data analysis component to extract semantic information from data, and a decision-making system that co-operates with a simulation module to automatically infer decisions of enhanced confidence. The decisions concern the risk of infectious disease spread in space and time, as well as the actions that need to be taken for disease control. Acknowledging that the population of the passengers accommodated in cruise ships is diverse, this chapter introduced also the CiL concept. CiL considers a personalized approach for the crew to contact passengers before taking any actions for disease control. The architecture of the presented *closed-loop* model is scalable, in the sense that it can be used to cover one or a few areas, or even a whole cruise ship.

A broad literature review was performed to identify alternatives for the implementation of the different modules comprising the closed-loop model, as well as open issues and relevant research challenges. Overall, the results of this study highlight the importance of the topic investigated in this chapter not only due to its socioeconomic impact, but also due to its potential scientific impact.

To develop effective systems for disease monitoring and control in cruise ships, one should consider the advantages and shortcomings that the selection and integration of different types of smart technologies entails. Overall, the combination of computational modeling approaches with various monitoring and decision-making technologies could further improve the development of effective control measures. It is worth noting that current disease spread monitoring methods primarily focus on large-scale prediction models in scenarios related to disease spreading in cities/countries, whereas fewer have a more local focus, such as cruise ships. Various RA tools have been applied to assist on decision making during epidemic outbreaks. Studies that concern the use of RA methods for epidemic outbreaks in ships have been limited, and they have been mainly based on knowledge-based systems. Most indoors-oriented RA tools have relied on mechanistic or stochastic epidemiological models. Moreover, there have been several studies suggesting guidelines and control measures for risk management of epidemic outbreaks in ships; however, developing a system that could assist in the application of such guidelines and measures, is still an open issue.

A variety of simulation models have been applied in the context of modeling disease spread in closed environments; nevertheless, most of them have focused on airborne pathogen transmission. Fomite-mediated transmission models are the most challenging since they should consider detailed contamination mechanics, such as fingertip-surface transmission. Furthermore, only a few of these models have been applied in the context of disease spread prediction in cruise ships. Aiming toward the development of an online simulation-based prediction system, it is important to investigate approaches for improved time-efficiency. Mathematical and numerical modeling approaches are usually computationally complex, requiring longer execution times. ROM approaches based on deep learning and AEs can be viable alternatives to improve their performance. ROM approaches may also be required in the case of computationally complex ABM simulations, where PINNs have been found to be promising solutions for this purpose.

The findings of this study can be used as a basis for the development of future systems for disease spread monitoring and control in ships and other relevant application areas. Research perspectives may extend even further, e.g., exploiting the closed-loop model as a basis for ship design.

Acknowledgments This chapter is supported by the European Union's Horizon Europe Research and Innovation Actions programme under grant agreement No 101069937, project name: HS4U (HEALTHY SHIP 4U). Views and opinions expressed are those of the author(s) only and do not necessarily reflect those of the European Union or the European Climate, Infrastructure, and Environment Executive Agency. Neither the European Union nor the granting authority can be held responsible for them.

References

1. Batista, B., Dickenson, D., Gurski, K., et al.: Minimizing disease spread on a quarantined cruise ship: a model of COVID-19 with asymptomatic infections. Math. Biosci. **329**, 108442 (2020)
2. Moriarty, L.F., Plucinski, M.M., Marston, B.J., et al.: Public health responses to COVID-19 outbreaks on cruise ships—worldwide, February–March 2020. Morb. Mortal. Wkly Rep. **69**, 347 (2020)
3. Mouchtouri, V.A., Koureas, M., Kyritsi, M., et al.: Environmental contamination of SARS-CoV-2 on surfaces, air-conditioner and ventilation systems. Int. J. Hyg. Environ. Health. **230**, 113599 (2020)
4. McCarter, Y.S.: Infectious disease outbreaks on cruise ships. Clin. Microbiol. Newsl. **31**, 161–168 (2009). https://doi.org/10.1016/j.clinmicnews.2009.10.001
5. Hill, C.D.: Cruise ship travel. In: Travel Medicine, pp. 377–382. Elsevier, U.S.A (2019)
6. Delikhoon, M., Guzman, M., Nabizadeh, R., et al.: Modes of transmission of severe acute respiratory syndrome-coronavirus-2 (SARS-CoV-2) and factors influencing on the airborne transmission: a review. Int J Environ Res Public Health. **18**, 395 (2021)
7. Bourouiba, L.: The fluid dynamics of disease transmission. Annu. Rev. Fluid Mech. **53**, 473–508 (2021)
8. Carlson, K.M., Boczek, L.A., Chae, S., Ryu, H.: Legionellosis and recent advances in technologies for Legionella control in premise plumbing systems: a review. Water. **12**, 676 (2020)
9. Leslie, E., Hinds, J., Hai, F.I.: Causes, factors, and control measures of opportunistic premise plumbing pathogens—a critical review. Appl. Sci. **11**, 4474 (2021)
10. Muli R, Tomi J, others (2020) Supplying ships with safe drinking-water. Int. Marit. Health 71:123–128
11. Prussin II, A.J., Schwake, D.O., Marr, L.C.: Ten questions concerning the aerosolization and transmission of Legionella in the built environment. Build. Environ. **123**, 684–695 (2017)
12. Pitol, A.K., Julian, T.R.: Community transmission of SARS-CoV-2 by surfaces: risks and risk reduction strategies. Environ. Sci. Technol. Lett. **8**, 263–269 (2021)
13. Chen, T.: Fomites and the COVID-19 pandemic: an evidence review on its role in viral transmission, vol. 19, p. 1953. National Collaborating Centre for Environmental Health, Vancouver (2021)
14. WHO, others: Handbook for Management of Public Health Events on Board Ships. WHO, Geneva (2016)
15. World Health Organization, others: International Medical Guide for Ships, 3rd edn. WHO, Geneva
16. Niku, S.B.: Introduction to Robotics: Analysis, Control, Applications. Wiley, U.S.A (2020)
17. Torra, V., Narukawa, Y.: Modeling Decisions: Information Fusion and Aggregation Operators. Springer Science & Business Media, Germany (2007)

18. Maeda, H., Sando, E., Toizumi, M., et al.: Epidemiology of coronavirus disease outbreak among crewmembers on cruise ship, Nagasaki City, Japan, April 2020. Emerg. Infect. Dis. **27**, 2251 (2021)
19. Fong, B., Fong, A., Li, C.: Infectious disease and indoor air quality management in a cruise ship environment. In: Smart Ships, pp. 101–116. CRC Press, Boca Raton, U.S.A (2022)
20. Mihai, V., Rusu, L.: An overview of the ship ventilation systems and measures to avoid the spread of diseases. Inventions. **6**, 55 (2021)
21. Nolich, M., Spoladore, D., Carciotti, S., et al.: Cabin as a home: a novel comfort optimization framework for IoT equipped smart environments and applications on cruise ships. Sensors. **19**, 1060 (2019)
22. Sága, Jr M., Bartoš, M., Zajacko, I., et al: An automated diagnostic and surveillance system for eliminating the community spread of infectious respiratory diseases in the industry. In: Advances in Design, Simulation and Manufacturing V: Proceedings of the 5th International Conference on Design, Simulation, Manufacturing: The Innovation Exchange, DSMIE-2022, June 7–10, 2022, Poznan, Poland–Volume 1: Manufacturing and Materials Engineering. Springer, pp 94–103 (2022)
23. Baig, M.M., GholamHosseini, H., Moqeem, A.A., et al.: A systematic review of wearable patient monitoring systems–current challenges and opportunities for clinical adoption. J. Med. Syst. **41**, 1–9 (2017)
24. Hoffman, J.S., Hirano, M., Panpradist, N., et al.: Passively sensing SARS-CoV-2 RNA in public transit buses. Sci. Total Environ. **821**, 152790 (2022)
25. Brewster, R.K., Chan, K., Allen, H., et al.: Future directions of infection control and risk management on military vessels: a narrative review. J. Public Health Emerg. **6**, 33 (2022)
26. Morawska, L., Tang, J.W., Bahnfleth, W., et al.: How can airborne transmission of COVID-19 indoors be minimised? Environ. Int. **142**, 105832 (2020)
27. Vardoulakis, S., Oyarce, D.A.E., Donner, E.: Transmission of COVID-19 and other infectious diseases in public washrooms: a systematic review. Sci. Total Environ. **803**, 149932 (2022)
28. Brewster, R.K., Sundermann, A., Boles, C.: Lessons learned for COVID-19 in the cruise ship industry. Toxicol. Ind. Health. **36**, 728–735 (2020). https://doi.org/10.1177/0748233720964631
29. Zhang, W., Xie, J., Gong, N., et al.: COVID-19 outbreaks on ships: analysis of three representative cases. Public Health Pract. **4**, 100320 (2022)
30. Aouad, S., Maizate, A., Zakari, A., Yassine, S.: A comprehensive survey of smart city technologies for monitoring and controlling the epidemic spread of COVID-19. In: Proceedings of the 4th International Conference on Networking, Information Systems & Security. pp 1–6 (2021)
31. Meraj, M., Alvi, S.A.M., Quasim, M.T., Haidar, S.W.: A critical review of detection and prediction of infectious disease using IOT sensors. In: 2021 second international conference on electronics and sustainable communication systems (ICESC). IEEE, pp 679–684 (2021)
32. Guk, K., Han, G., Lim, J., et al.: Evolution of wearable devices with real-time disease monitoring for personalized healthcare. Nanomaterials. **9**, 813 (2019)
33. Vairavan, M.: Thermal camera-based COVID-19 detection. In: Proceedings of Third International Conference on Intelligent Computing, Information and Control Systems: ICICCS 2021. Springer, pp 395–402 (2022)
34. Salman, N., Kemp, A.H., Khan, A., Noakes, C.: Real time wireless sensor network (WSN) based indoor air quality monitoring system. IFAC-PapersOnLine. **52**, 324–327 (2019)
35. Mohammed, M., Syamsudin, H., Al-Zubaidi, S., et al.: Novel COVID-19 detection and diagnosis system using IOT based smart helmet. Int. J. Psychosoc. Rehabil. **24**, 2296–2303 (2020)
36. Jain, S., Nehra, M., Kumar, R., et al.: Internet of medical things (IoMT)-integrated biosensors for point-of-care testing of infectious diseases. Biosens. Bioelectron. **179**, 113074 (2021)
37. Pal, A., Sankarasubbu, M.: Pay attention to the cough: early diagnosis of COVID-19 using interpretable symptoms embeddings with cough sound signal processing. In: Proceedings of the 36th Annual ACM Symposium on Applied Computing. pp 620–628 (2021)

38. Aly, M., Rahouma, K.H., Ramzy, S.M.: Pay attention to the speech: COVID-19 diagnosis using machine learning and crowdsourced respiratory and speech recordings. Alex. Eng. J. **61**, 3487–3500 (2022)
39. Zhang, H., Srinivasan, R., Ganesan, V.: Low cost, multi-pollutant sensing system using raspberry Pi for indoor air quality monitoring. Sustainability. **13**, 370 (2021)
40. Kanál, A.K., Tamás, K.: Assessment of indoor air quality of educational facilities using an IoT solution for a healthy learning environment. In: 2020 IEEE International Instrumentation and Measurement Technology Conference (I2MTC). pp 1–6 (2020)
41. Ameer, S., Shah, M.A., Khan, A., et al.: Comparative analysis of machine learning techniques for predicting air quality in smart cities. IEEE Access. **7**, 128325–128338 (2019). https://doi.org/10.1109/ACCESS.2019.2925082
42. Roskams, M., Haynes, B.: Predictive analytics in facilities management: a pilot study for predicting environmental comfort using wireless sensors. J. Facil. Manag. **17**, 356–370 (2019)
43. Awotunde, J.B., Jimoh, R.G., Folorunso, S.O., et al.: Privacy and security concerns in IoT-based healthcare systems. In: The Fusion of Internet of Things, Artificial Intelligence, and Cloud Computing in Health Care, pp. 105–134. Springer, Switzerland (2021)
44. Yan, S., Chughtai, A., Macintyre, C.: Utility and potential of rapid epidemic intelligence from internet-based sources. Int. J. Infect. Dis. **63**, 77–87 (2017)
45. Hossain, M.S., Muhammad, G., Guizani, N.: Explainable AI and mass surveillance system-based healthcare framework to combat COVID-I9 like pandemics. IEEE Netw. **34**, 126–132 (2020)
46. Voigt, P., Von dem Bussche, A.: The EU General Data Protection Regulation (GDPR). A Practical Guide, vol. 10, 1st edn, pp. 10–5555. Springer International Publishing, Cham (2017)
47. World Health Organization, others: Rapid Risk Assessment of Acute Public Health Events. World Health Organization, Geneva (2012)
48. Han, X.: Review of risk assessment methods for infectious diseases. In: IOP Conference Series: Earth and Environmental Science, vol. 446, p. 032023. IOP Publishing, Chongqing, China (2020)
49. Faustman, E.M., Omenn, G.S.: Risk assessment. In: Casarett and Doull's Toxicology: The Basic Science of Poisons, 8th edn. McGraw-Hill Education, New York (2012)
50. Ross, Y.B., Hoque, M., Blanton, J.D., et al.: Rabies healthcare-seeking behaviors of urban and peri-urban residents: results from a rabies knowledge, attitudes, and practices survey, Bangladesh, 2018. PLoS Negl. Trop. Dis. **16**, e0010634 (2022)
51. Ewing, M., Naredi, P., Zhang, C., Månsson, J.: Identification of patients with non-metastatic colorectal cancer in primary care: a case-control study. Br. J. Gen. Pract. **66**, e880–e886 (2016)
52. Moynihan, R., Sanders, S., Michaleff, Z.A., et al.: Impact of COVID-19 pandemic on utilisation of healthcare services: a systematic review. BMJ Open. **11**, e045343 (2021)
53. Xiao, H., Dai, X., Wagenaar, B.H., et al.: The impact of the COVID-19 pandemic on health services utilization in China: time-series analyses for 2016–2020. Lancet Reg. Health West. Pac. **9**, 100122 (2021)
54. Recchia, V., Aloisi, A., Zizza, A.: Risk management and communication plans from SARS to COVID-19 and beyond. Int. J. Health Plann. Manag. **37**, 3039–3060 (2022)
55. Calò, F., Russo, A., Camaioni, C., et al.: Burden, risk assessment, surveillance and management of SARS-CoV-2 infection in health workers: a scoping review. Infect. Dis. Poverty. **9**, 1–11 (2020)
56. Bertagna, S., Dodero, M., Bortuzzo, V., et al.: An innovative approach for the biological risk management on-board ships during COVID-19 crisis. Marit. Transport Res. **2**, 100028 (2021)
57. Kordsmeyer, A.-C., Mojtahedzadeh, N., Heidrich, J., et al.: Systematic review on outbreaks of SARS-CoV-2 on cruise, navy and cargo ships. Int. J. Environ. Res. Public Health. **18**, 5195 (2021)

58. Huang, L.-S., Li, L., Dunn, L., He, M.: Taking account of asymptomatic infections: a modeling study of the COVID-19 outbreak on the Diamond Princess cruise ship. PLoS One. **16**, e0248273 (2021)
59. Willebrand, K.S., Pischel, L., Malik, A.A., et al.: A review of COVID-19 transmission dynamics and clinical outcomes on cruise ships worldwide, January to October 2020. Eur. Secur. **27**, 2002113 (2022)
60. Rooney, R.M., Bartram, J.K., Cramer, E.H., et al.: A review of outbreaks of waterborne disease associated with ships: evidence for risk management. Public Health Rep. **119**, 435–442 (2004)
61. Guagliardo, S.A.J., Prasad, P.V., Rodriguez, A., et al.: Cruise ship travel in the era of coronavirus disease 2019 (COVID-19): a summary of outbreaks and a model of public health interventions. Clin. Infect. Dis. **74**, 490–497 (2022)
62. Wang, Z., Yao, M., Meng, C., Claramunt, C.: Risk assessment of the overseas imported COVID-19 of ocean-going ships based on AIS and infection data. ISPRS Int. J. Geo Inf. **9**, 351 (2020)
63. Ventikos, N.P., Sotiralis, P., Annetis, M., Roland, F.: Developing a framework for health risk assessment, by integrating infection and spreading aspects into RBD. In: Medical Sciences Forum, p. 2. Multidisciplinary Digital Publishing Institute, Athens, Greece (2022)
64. Braidotti, L., Bertagna, S., Dodero, M., et al.: Identification of measures to contain the outbreaks on passenger ships using pedestrian simulations. Procedia Comput. Sci. **200**, 1565–1574 (2022)
65. Pirneskoski, J., Kuisma, M., Olkkola, K.T., Nurmi, J.: Prehospital national early warning score predicts early mortality. Acta Anaesthesiol. Scand. **63**, 676–683 (2019)
66. Duan, J., Jiao, F.: Novel case-based reasoning system for public health emergencies. Risk Manag. Healthc. Policy. **14**, 541–553 (2021)
67. Ling, M., Wenxiao, T., Yenan, F., et al.: Risk assessment of public health emergencies concerned in the mainland of China, March 2021. Dis. Surveill. **36**, 201–203 (2021)
68. Wen-yan, X.: Establishment and application on risk assessment index system of imported respiratory infectious diseases at frontier ports (2012)
69. World Health Organization, others: Tool for Influenza Pandemic Risk Assessment (TIPRA). World Health Organization, Geneva (2016)
70. Soltanzadeh, A., Adeli, S.H., Yarandi, M.S., Mahdinia, M.: Developing a risk assessment method for infectious diseases focusing on COVID-19 based on the Delphi method and fuzzy analytical hierarchy process. Work, 1–10 (2022)
71. Duan, J., Lin, Z., Jiao, F., et al.: A dynamic case-based reasoning system for responding to infectious disease outbreaks. Expert Syst. Appl. **204**, 117628 (2022)
72. Schröder, I.: COVID-19: a risk assessment perspective. ACS Chem. Health Saf. **27**, 160–169 (2020). https://doi.org/10.1021/acs.chas.0c00035
73. Zhou, S., Zhou, S., Zheng, Z., et al.: Risk assessment for precise intervention of COVID-19 epidemic based on available big data and spatio-temporal simulation method: empirical evidence from different public places in Guangzhou, China. Appl. Geogr. **143**, 102702 (2022)
74. Moss, R., Hickson, R.I., McVernon, J., et al.: Model-informed risk assessment and decision making for an emerging infectious disease in the Asia-Pacific region. PLoS Negl. Trop. Dis. **10**, e0005018 (2016)
75. Lesmanawati, D.A., Veenstra, P., Moa, A., et al.: A rapid risk analysis tool to prioritise response to infectious disease outbreaks. BMJ Glob. Health. **5**, e002327 (2020)
76. Zhou, L., Liu, J.-M., Dong, X.-P., et al.: COVID-19 seeding time and doubling time model: an early epidemic risk assessment tool. Infect. Dis. Poverty. **9**, 1–9 (2020)
77. Chatterjee, R., Bajwa, S., Dwivedi, D., et al.: COVID-19 risk assessment tool: dual application of risk communication and risk governance. Prog. Disaster Sci. **7**, 100109 (2020)
78. Jia, J.S., Lu, X., Yuan, Y., et al.: Population flow drives spatio-temporal distribution of COVID-19 in China. Nature. **582**, 389–394 (2020)
79. Pluchino, A., Biondo, A., Giuffrida, N., et al.: A novel methodology for epidemic risk assessment of COVID-19 outbreak. Sci. Rep. **11**, 1–20 (2021)

80. Helou, M.A., DiazGranados, D., Ryan, M.S., Cyrus, J.W.: Uncertainty in decision-making in medicine: a scoping review and thematic analysis of conceptual models. Acad. Med. **95**, 157 (2020)
81. Begoli, E., Bhattacharya, T., Kusnezov, D.: The need for uncertainty quantification in machine-assisted medical decision making. Nat. Mach. Intell. **1**, 20–23 (2019)
82. Rasmussen, S.L., Sahay, S.: Engaging with uncertainty: information practices in the context of disease surveillance in Burkina Faso. Inf. Organ. **31**, 100366 (2021)
83. Arji, G., Ahmadi, H., Nilashi, M., et al.: Fuzzy logic approach for infectious disease diagnosis: a methodical evaluation, literature and classification. Biocybern. Biomed. Eng. **39**, 937–955 (2019)
84. Jiang, Q., Zhou, X., Wang, R., et al.: Intelligent monitoring for infectious diseases with fuzzy systems and edge computing: a survey. Appl. Soft Comput. **108835**, 108835 (2022)
85. Seker, S., Baglan, F.B., Aydin, N., et al.: Risk assessment approach for analyzing risk factors to overcome pandemic using interval-valued q-rung orthopair fuzzy decision making method. Appl. Soft Comput. **132**, 109891 (2023)
86. Painuli, D., Mishra, D., Bhardwaj, S., Aggarwal, M.: Fuzzy rule based system to predict COVID19-a deadly virus. Way. **3**, 5 (2020)
87. Groumpos, P.P., Apostolopoulos, I.D.: Modeling the spread of dangerous pandemics with the utilization of a hybrid-statistical–Advanced-Fuzzy-Cognitive-Map algorithm: the example of COVID-19. Res. Biomed. Eng. **37**, 749–764 (2021)
88. Crabtree, K., Gerba, C., Rose, J., Haas, C.: Waterborne adenovirus: a risk assessment. Water Sci. Technol. **35**, 1–6 (1997)
89. Pintar, K., Fazil, A., Pollari, F., et al.: A risk assessment model to evaluate the role of fecal contamination in recreational water on the incidence of cryptosporidiosis at the community level in Ontario. Risk Anal. **30**, 49–64 (2010)
90. Jacob, P., Henry, A., Meheut, G., et al.: Health risk assessment related to waterborne pathogens from the river to the tap. Int. J. Environ. Res. Public Health. **12**, 2967–2983 (2015)
91. Sims, N., Kasprzyk-Hordern, B.: Future perspectives of wastewater-based epidemiology: monitoring infectious disease spread and resistance to the community level. Environ. Int. **139**, 105689 (2020)
92. Sze To, G.N., Chao, C.Y.H.: Review and comparison between the Wells–Riley and dose-response approaches to risk assessment of infectious respiratory diseases. Indoor Air. **20**, 2–16 (2010)
93. Azimi, P., Keshavarz, Z., Cedeno Laurent, J.G., et al.: Mechanistic transmission modeling of COVID-19 on the Diamond Princess cruise ship demonstrates the importance of aerosol transmission. Proc. Natl. Acad. Sci. **118**, e2015482118 (2021)
94. Wan, P.K., Huang, L., Lai, Z., et al.: Automated infection risks assessments (AIRa) for decision-making using a blockchain-based alert system: a case study in a representative building. Environ. Res. **216**, 114663 (2023)
95. Di Gilio, A., Palmisani, J., Pulimeno, M., et al.: CO2 concentration monitoring inside educational buildings as a strategic tool to reduce the risk of Sars-CoV-2 airborne transmission. Environ. Res. **202**, 111560 (2021)
96. Russell, T.W., Hellewell, J., Jarvis, C.I., et al.: Estimating the infection and case fatality ratio for coronavirus disease (COVID-19) using age-adjusted data from the outbreak on the Diamond Princess cruise ship, February 2020. Eurosurveillance. **25**, 2000256 (2020)
97. Mizumoto, K., Kagaya, K., Zarebski, A., Chowell, G.: Estimating the asymptomatic proportion of coronavirus disease 2019 (COVID-19) cases on board the Diamond Princess cruise ship, Yokohama, Japan, 2020. Eurosurveillance. **25**, 2000180 (2020)
98. Zhang, S., Diao, M., Yu, W., et al.: Estimation of the reproductive number of novel coronavirus (COVID-19) and the probable outbreak size on the Diamond Princess cruise ship: a data-driven analysis. Int. J. Infect. Dis. **93**, 201–204 (2020)
99. Mizumoto, K., Chowell, G.: Transmission potential of the novel coronavirus (COVID-19) onboard the diamond Princess Cruises Ship, 2020. Infect. Dis. Model. **5**, 264–270 (2020)

100. Roth, J.A., Battegay, M., Juchler, F., et al.: Introduction to machine learning in digital healthcare epidemiology. Infect. Control Hosp. Epidemiol. **39**, 1457–1462 (2018)
101. He, X., Zhou, C., Wang, Y., Yuan, X.: Risk assessment and prediction of COVID-19 based on epidemiological data from spatiotemporal geography. Front. Environ. Sci. **9**, 634156 (2021)
102. Nikparvar, B., Rahman, M.M., Hatami, F., Thill, J.-C.: Spatio-temporal prediction of the COVID-19 pandemic in US counties: modeling with a deep LSTM neural network. Sci. Rep. **11**, 21715 (2021)
103. Wang, P., Liu, H., Zheng, X., Ma, R.: A new method for spatio-temporal transmission prediction of COVID-19. Chaos, Solitons Fractals. **167**, 112996 (2023)
104. Paul, S.K., Jana, S., Bhaumik, P.: A multivariate spatiotemporal model of COVID-19 epidemic using ensemble of ConvLSTM networks. J. Inst. Eng. India Ser. B. **102**, 1137–1142 (2021)
105. Niraula, P., Mateu, J., Chaudhuri, S.: A Bayesian machine learning approach for spatio-temporal prediction of COVID-19 cases. Stoch. Env. Res. Risk A. **36**, 2265–2283 (2022)
106. Padmanabhan, R., Meskin, N., Khattab, T., et al.: Reinforcement learning-based decision support system for COVID-19. Biomed. Signal Process. Control. **68**, 102676 (2021)
107. Guo, X., Chen, P., Liang, S., et al.: PaCAR: COVID-19 pandemic control decision making via large-scale agent-based modeling and deep reinforcement learning. Med. Decis. Mak. **42**, 1064–1077 (2022)
108. Spangler, D., Hermansson, T., Smekal, D., Blomberg, H.: A validation of machine learning-based risk scores in the prehospital setting. PLoS One. **14**, e0226518 (2019)
109. Wang, L., Guo, J., Tian, Z., et al.: Developing a hybrid risk assessment tool for familial hypercholesterolemia: a machine learning study of Chinese arteriosclerotic cardiovascular disease patients. Front. Cardiovasc. Med. **9** (2022)
110. Chung, K., Jung, H.: Knowledge-based dynamic cluster model for healthcare management using a convolutional neural network. Inf. Technol. Manag. **21**, 41–50 (2020)
111. Soui, M., Gasmi, I., Smiti, S., Ghédira, K.: Rule-based credit risk assessment model using multi-objective evolutionary algorithms. Expert Syst. Appl. **126**, 144–157 (2019)
112. Liu, M., Liu, M., Li, Z., et al.: The spatial clustering analysis of COVID-19 and its associated factors in mainland China at the prefecture level. Sci. Total Environ. **777**, 145992 (2021)
113. Younsi, F.-Z., Bounnekar, A., Hamdadou, D., Boussaid, O.: Integration of multiple regression model in an epidemiological decision support system. Int. J. Inf. Technol. Decis. Making. **18**, 1755–1783 (2019)
114. Ghosh, D., Guha, R.: Using a neural network for mining interpretable relationships of West Nile risk factors. Soc. Sci. Med. **72**, 418–429 (2011)
115. Tian, T., Jiang, Y., Zhang, Y., et al.: COVID-Net: a deep learning based and interpretable predication model for the county-wise trajectories of COVID-19 in the United States. MedRxiv, 2020–2025 (2020)
116. Chakraborty, T., Ghosh, I.: Real-time forecasts and risk assessment of novel coronavirus (COVID-19) cases: a data-driven analysis. Chaos, Solitons Fractals. **135**, 109850 (2020)
117. Sykes, A.L., Silva, G.S., Holtkamp, D.J., et al.: Interpretable machine learning applied to on-farm biosecurity and porcine reproductive and respiratory syndrome virus. Transbound. Emerg. Dis. **69**, e916–e930 (2022)
118. Bazilevych, K., Chumachenko, D., Hulianytskyi, L., et al.: Intelligent decision-support system for epidemiological diagnostics. I. A concept of architecture design. Cybern. Syst. Anal. **58**, 343–353 (2022)
119. Deng, Q., Xiao, X., Zhu, L., et al.: A national risk analysis model (NRAM) for the assessment of COVID-19 epidemic. Risk Anal. **43**, 1946 (2023)
120. Kumar, R., Al-Turjman, F., Srinivas, L., et al.: ANFIS for prediction of epidemic peak and infected cases for COVID-19 in India. Neural Comput. & Applic., 1–14 (2021)
121. Alsayed, A., Sadir, H., Kamil, R., Sari, H.: Prediction of epidemic peak and infected cases for COVID-19 disease in Malaysia, 2020. Int. J. Environ. Res. Public Health. **17**, 4076 (2020)
122. Cihan, P.: Fuzzy rule-based system for predicting daily case in covid-19 outbreak. In: 2020 4th International Symposium on Multidisciplinary Studies and Innovative Technologies (ISMSIT). IEEE, pp 1–4 (2020)

123. Papageorgiou, E.I., Papandrianos, N.I., Karagianni, G., et al.: A fuzzy cognitive map based tool for prediction of infectious diseases. In: 2009 IEEE International Conference on Fuzzy Systems. IEEE, pp 2094–2099 (2009)
124. Li, H., Meng, S., Tong, H.: How to control cruise ship disease risk? Inspiration from the research literature. Mar. Policy. **132**, 104652 (2021)
125. Joob, B., Wiwanitkit, V.: Concern on infectious disease prevention: an issue for maritime management. Maritime Technol. Res. **5**, 262682 (2023)
126. Lipinski, T., Ahmad, D., Serey, N., Jouhara, H.: Review of ventilation strategies to reduce the risk of disease transmission in high occupancy buildings. Int. J. Thermofluids. **7**, 100045 (2020)
127. Almilaji, O.: Air recirculation role in the spread of COVID-19 onboard the diamond princess cruise ship during a quarantine period. Aerosol Air Qual. Res. **21**, 200495 (2021)
128. Gupta, A., Kunte, R., Goyal, N., et al.: A comparative analysis of control measures on-board ship against COVID-19 and similar novel viral respiratory disease outbreak: quarantine ship or disembark suspects? Med. J. Armed Forces India. **77**, S430–S436 (2021)
129. Hadjichristodoulou, C., Mouchtouri, V.A., Guglielmetti, P., et al.: Actions for prevention and control of health threats related to maritime transport in European Union. Travel Med. Infect. Dis. **11**, 238–242 (2013)
130. Laganà, P., Gambuzza, M.E., Delia, S.: Legionella risk assessment in cruise ships and ferries. Ann. Agric. Environ. Med. **24**, 276 (2017)
131. Gollasch, S., Minchin, D., David, M.: The transfer of harmful aquatic organisms and pathogens with ballast water and their impacts. In: Global Maritime Transport and Ballast Water Management: Issues and Solutions, pp. 35–58. Springer, Dordrecht (2015)
132. Xu, L., Peng, J., Wang, M., Yang, J.: Lessons and suggestions to travelers and cruise ships in the fight against COVID-19. QJM. **114**, 153–154 (2021)
133. Codreanu, T.A., Ngeh, S., Trewin, A., Armstrong, P.K.: Successful control of an onboard COVID-19 outbreak using the cruise ship as a quarantine facility, Western Australia, Australia. Emerg. Infect. Dis. **27**, 1279 (2021)
134. Hatzianastasiou, S., Mouchtouri, V.A., Pavli, A., et al.: COVID-19 outbreak on a passenger ship and assessment of response measures, Greece, 2020. Emerg. Infect. Dis. **27**, 1927 (2021)
135. Walker, L.J., Codreanu, T.A., Armstrong, P.K., et al.: SARS-CoV-2 infections among Australian passengers on the Diamond Princess cruise ship: a retrospective cohort study. PLoS One. **16**, e0255401 (2021)
136. Peng, S., Chen, Q., Liu, E.: The role of computational fluid dynamics tools on investigation of pathogen transmission: prevention and control. Sci. Total Environ. **746**, 142090 (2020). https://doi.org/10.1016/j.scitotenv.2020.142090
137. Li, Y., Wang, J.-X., Chen, X.: Can a toilet promote virus transmission? From a fluid dynamics perspective. Phys. Fluids. **32**, 065107 (2020). https://doi.org/10.1063/5.0013318
138. Wang, J.-X., Li, Y.-Y., Liu, X.-D., Cao, X.: Virus transmission from urinals. Phys. Fluids. **32**, 081703 (2020). https://doi.org/10.1063/5.0021450
139. Cao, X., Hao, G., Li, Y., et al.: On male urination and related environmental disease transmission in restrooms: from the perspectives of fluid dynamics. Sustain. Cities Soc. **80**, 103753 (2022). https://doi.org/10.1016/j.scs.2022.103753
140. Canales, R.A., Reynolds, K.A., Wilson, A.M., et al.: Modeling the role of fomites in a norovirus outbreak. J. Occup. Environ. Hyg. **16**, 16–26 (2019)
141. Wilson, A.M., King, M.-F., López-Garcıa, M., et al.: Evaluating a transfer gradient assumption in a fomite-mediated microbial transmission model using an experimental and Bayesian approach. J. R. Soc. Interface. **17**, 20200121 (2020)
142. Zhang, W., Liu, S., Osgood, N., et al.: Using simulation modelling and systems science to help contain COVID-19: a systematic review. Syst. Res. Behav. Sci. **40**, 207–234 (2023)
143. Ghoroghi, A., Rezgui, Y., Wallace, R.: Impact of ventilation and avoidance measures on SARS-CoV-2 risk of infection in public indoor environments. Sci. Total Environ. **838**, 156518 (2022)

144. Harweg, T., Wagner, M., Weichert, F.: Agent-based simulation for infectious disease modelling over a period of multiple days, with application to an airport scenario. Int. J. Environ. Res. Public Health. **20**, 545 (2023)
145. Yin, S., Sze-To, G., Chao, C.Y.: Retrospective analysis of multi-drug resistant tuberculosis outbreak during a flight using computational fluid dynamics and infection risk assessment. Build. Environ. **47**, 50–57 (2012)
146. La, A., Zhang, Q., Cicek, N., et al.: Dose–response modelling of infectious animal diseases coupled with computational fluid dynamics: a simulation of airborne porcine reproductive and respiratory syndrome virus. Biosyst. Eng. **208**, 58–78 (2021)
147. Gupta, J.K., Lin, C.-H., Chen, Q.: Risk assessment of airborne infectious diseases in aircraft cabins. Indoor Air. **22**, 388–395 (2012)
148. Qian, H., Li, Y., Nielsen, P.V., Huang, X.: Spatial distribution of infection risk of SARS transmission in a hospital ward. Build. Environ. **44**, 1651–1658 (2009)
149. Xiao, S., Li, Y., Sung, M., et al.: A study of the probable transmission routes of MERS-CoV during the first hospital outbreak in the Republic of Korea. Indoor Air. **28**, 51–63 (2017). https://doi.org/10.1111/ina.12430
150. Tang, L., Zhou, Y., Wang, L., et al.: A review of multi-compartment infectious disease models. Int. Stat. Rev. **88**, 462–513 (2020). https://doi.org/10.1111/insr.12402
151. Zhao, X., Liu, S., Yin, Y., et al.: Airborne transmission of COVID-19 virus in enclosed spaces: an overview of research methods. Indoor Air. **32**, e13056 (2022)
152. Burridge, H.C., Bhagat, R.K., Stettler, M.E., et al.: The ventilation of buildings and other mitigating measures for COVID-19: a focus on wintertime. Proc. R. Soc. A. **477**, 20200855 (2021)
153. Miller, S.L., Nazaroff, W.W., Jimenez, J.L., et al.: Transmission of SARS-CoV-2 by inhalation of respiratory aerosol in the Skagit Valley Chorale superspreading event. Indoor Air. **31**, 314–323 (2021)
154. Rowe, B.R., Canosa, A., Drouffe, J.-M., Mitchell, J.B.A.: Simple quantitative assessment of the outdoor versus indoor airborne transmission of viruses and COVID-19. Environ. Res. **198**, 111189 (2021)
155. Wiryasaputra, R., Huang, C.-Y., Kristiani, E., et al.: Review of an intelligent indoor environment monitoring and management system for COVID-19 risk mitigation. Front. Public Health. **10** (2022)
156. Baboli, Z., Neisi, N., Babaei, A.A., et al.: On the airborne transmission of SARS-CoV-2 and relationship with indoor conditions at a hospital. Atmos. Environ. **261**, 118563 (2021)
157. Kurabuchi, T., Yanagi, U., Ogata, M., et al.: Operation of air-conditioning and sanitary equipment for SARS-CoV-2 infectious disease control. Jpn Architect. Rev. **4**, 608–620 (2021)
158. Megahed, N.A., Ghoneim, E.M.: Indoor air quality: rethinking rules of building design strategies in post-pandemic architecture. Environ. Res. **193**, 110471 (2021)
159. Qian, H., Zheng, X.: Ventilation control for airborne transmission of human exhaled bioaerosols in buildings. J. Thorac. Dis. **10**, S2295 (2018)
160. Hussain, S.M., Goel, S., Kadapa, C., Aristodemou, E.: A short review of vapour droplet dispersion models used in CFD to study the airborne spread of COVID19. Materials Today: Proceedings (2022)
161. Liu, H., He, S., Shen, L., Hong, J.: Simulation-based study of COVID-19 outbreak associated with air-conditioning in a restaurant. Phys. Fluids. **33**, 023301 (2021)
162. Dbouk, T., Drikakis, D.: On coughing and airborne droplet transmission to humans. Phys. Fluids. **32**, 053310 (2020). https://doi.org/10.1063/5.0011960
163. Ho, C.K.: Modeling airborne pathogen transport and transmission risks of SARS-CoV-2. Appl. Math. Model. **95**, 297–319 (2021)
164. Dbouk, T., Drikakis, D.: On respiratory droplets and face masks. Phys. Fluids. (2020). https://doi.org/10.1063/5.0015044
165. Dbouk, T., Drikakis, D.: On airborne virus transmission in elevators and confined spaces. Phys. Fluids. (2021). https://doi.org/10.1063/5.0038180

166. Li, Y., Qian, H., Hang, J., et al.: Probable airborne transmission of SARS-CoV-2 in a poorly ventilated restaurant. Build. Environ. **196**, 107788 (2021)
167. Abuhegazy, M., Talaat, K., Anderoglu, O., et al.: Numerical investigation of aerosol transport in a classroom with relevance to COVID-19. Physics. **32**, 103311 (2020). https://doi.org/10.1063/5.0029118
168. Yang, X., Ou, C., Yang, H., et al.: Transmission of pathogen-laden expiratory droplets in a coach bus. J. Hazard. Mater. **397**, 122609 (2020)
169. Zhang, Z., Han, T., Yoo, K., et al.: Disease transmission through expiratory aerosols on an urban bus. Physics. **33**, 015116 (2021). https://doi.org/10.1063/5.0037452
170. D'Alicandro, A.C., Massarotti, N., Mauro, A.: Aerosol hazards in operating rooms: a review of numerical and experimental studies. J. Aerosol Sci. **158**, 105823 (2021)
171. Reveil, M., Chen, Y.-H.: Predicting and preventing COVID-19 outbreaks in indoor environments: an agent-based modeling study. Sci. Rep. **12**, 16076 (2022)
172. Löhner, R., Antil, H., Srinivasan, A., et al.: High-fidelity simulation of pathogen propagation, transmission and mitigation in the built environment. Arch. Comput. Methods Eng. **28**, 4237–4262 (2021)
173. Wang, J., Tang, H., Wang, J., Zhong, Z.: An agent-based study on the airborne transmission risk of infectious disease in a fever clinic during COVID-19 pandemic. Build. Environ. **218**, 109118 (2022)
174. Löhner, R., Antil, H., Gimenez, J.M., et al.: A deterministic pathogen transmission model based on high-fidelity physics. Comput. Methods Appl. Mech. Eng. **401**, 114929 (2022)
175. Altamimi, T., Khalil, H., Rajus, V.S., et al.: Cell-DEVS models with BIM integration for airborne transmission of COVID-19 indoors. In: Symposium on Simulation in Architecture and Urban Design SimAUD (2021)
176. Farthing, T.S., Lanzas, C.: When can we stop wearing masks? Agent-based modeling to identify when vaccine coverage makes nonpharmaceutical interventions for reducing SARS-CoV-2 infections redundant in indoor gatherings. MedRxiv. (2021). https://doi.org/10.1101/2021.04.19.21255737
177. Petropoulos, F., Makridakis, S., Stylianou, N.: COVID-19: forecasting confirmed cases and deaths with a simple time series model. Int. J. Forecast. **38**, 439–452 (2022)
178. Tsang, T.-W., Mui, K.-W., Wong, L.-T.: Computational Fluid Dynamics (CFD) studies on airborne transmission in hospitals: a review on the research approaches and the challenges. J. Build. Eng. **63**, 105533 (2022)
179. Guo, Y., Qian, H., Sun, Z., et al.: Assessing and controlling infection risk with Wells-Riley model and spatial flow impact factor (SFIF). Sustain. Cities Soc. **67**, 102719 (2021)
180. Motamedi, H., Shirzadi, M., Tominaga, Y., Mirzaei, P.A.: CFD modeling of airborne pathogen transmission of COVID-19 in confined spaces under different ventilation strategies. Sustain. Cities Soc. **76**, 103397 (2022)
181. Wan, M.P., Sze To, G.N., Chao, C.Y.H., et al.: Modeling the fate of expiratory aerosols and the associated infection risk in an aircraft cabin environment. Aerosol Sci. Technol. **43**, 322–343 (2009)
182. Zhang, Z., Capecelatro, J., Maki, K.: On the utility of a well-mixed model for predicting disease transmission on an urban bus. AIP Adv. **11**, 085229 (2021)
183. Chen, C., Lin, C.-H., Long, Z., Chen, Q.: Predicting transient particle transport in enclosed environments with the combined computational fluid dynamics and M arkov chain method. Indoor Air. **24**, 81–92 (2014)
184. Tan, S., Zhang, Z., Maki, K., et al.: Beyond well-mixed: a simple probabilistic model of airborne disease transmission in indoor spaces. Indoor Air. **32**, e13015 (2022)
185. Huang, J., Jones, P., Zhang, A., et al.: Outdoor airborne transmission of coronavirus among apartments in high-density cities. Front. Built Environ. **48** (2021)
186. Aliabadi, A.A., Rogak, S.N., Bartlett, K.H., Green, S.I.: Preventing airborne disease transmission: review of methods for ventilation design in health care facilities. Adv. Prev. Med. **2011**, 1 (2011)

187. Cheng, P., Luo, K., Xiao, S., et al.: Predominant airborne transmission and insignificant fomite transmission of SARS-CoV-2 in a two-bus COVID-19 outbreak originating from the same pre-symptomatic index case. J. Hazard. Mater. **425**, 128051 (2022)
188. Mukherjee, D., Wadhwa, G.: A mesoscale agent based modeling framework for flow-mediated infection transmission in indoor occupied spaces. Comput. Methods Appl. Mech. Eng. **401**, 115485 (2022)
189. Schinko, C., Shao, L., Mueller-Roemer, J., et al.: Accelerated airborne virus spread simulation: coupling agent-based modeling with GPU-accelerated computational fluid dynamics. In: VISIGRAPP (1: GRAPP). pp. 278–285 (2022)
190. Lazebnik, T., Alexi, A.: High resolution spatio-temporal model for room-level airborne pandemic spread. Mathematics. **11**, 426 (2023)
191. Marlow, F., Jacob, J., Sagaut, P.: A multidisciplinary model coupling Lattice-Boltzmann-based CFD and a Social Force Model for the simulation of pollutant dispersion in evacuation situations. Build. Environ. **205**, 108212 (2021)
192. Xiang, L., Lee, C.W., Zikanov, O., et al.: Reduced Order Modeling of Transport of Infectious Aerosols in Ventilated Rooms. Available at SSRN 4334331 (2023)
193. Rudnick, S., Milton, D.K.: Risk of indoor airborne infection transmission estimated from carbon dioxide concentration. Indoor Air. **13**, 237–245 (2003)
194. Xu, Q., Chraibi, M.: On the effectiveness of the measures in supermarkets for reducing contact among customers during COVID-19 period. Sustainability. **12**, 9385 (2020)
195. Carlotti, P., Massoulié, B., Morez, A., et al.: Respiratory pandemic and indoor aeraulics of classrooms. Build. Environ. **212**, 108756 (2022)
196. Wang, Q., Li, Y., Lung, D.C., et al.: Aerosol transmission of SARS-CoV-2 due to the chimney effect in two high-rise housing drainage stacks. J. Hazard. Mater. **421**, 126799 (2022)
197. Ali, B.H., Shahin, M.S., Sangani, M.M.M., et al.: Wastewater aerosols produced during flushing toilets, WWTPs, and irrigation with reclaimed municipal wastewater as indirect exposure to SARS-CoV-2. J. Environ. Chem. Eng. **9**, 106201 (2021)
198. Wang, J.-X., Wu, Z., Wang, H., et al.: Ventilation reconstruction in bathrooms for restraining hazardous plume: mitigate COVID-19 and beyond. J. Hazard. Mater. **439**, 129697 (2022). https://doi.org/10.1016/j.jhazmat.2022.129697
199. Zhang, L., Wang, Z.-C., Zhang, Y.: Dynamics of a reaction–diffusion waterborne pathogen model with direct and indirect transmission. Comput. Math. Appl. **72**, 202–215 (2016). https://doi.org/10.1016/j.camwa.2016.04.046
200. Heida, A., Mraz, A., Hamilton, M.T., et al.: Computational framework for evaluating risk trade-offs in costs associated with legionnaires' disease risk, energy, and scalding risk for hospital hot water systems. Environ. Sci. Water Res. Technol. **8**, 76–97 (2022)
201. Hamilton, K., Haas, C.: Critical review of mathematical approaches for quantitative microbial risk assessment (QMRA) of Legionella in engineered water systems: research gaps and a new framework. Environ. Sci. Water Res. Technol. **2**, 599–613 (2016)
202. Mari, L., Casagrandi, R., Bertuzzo, E., et al.: Conditions for transient epidemics of waterborne disease in spatially explicit systems. R. Soc. Open Sci. **6**, 181517 (2019)
203. Chaysiri, R., Louis, G.E., Chinviriyasit, W.: Modeling the health impact of water and sanitation service deficits on waterborne disease transmission. Adv. Differ. Equ. **2021**, 405 (2021)
204. Sen, N.: Transmission and evaporation of cough droplets in an elevator: numerical simulations of some possible scenarios. Phys. Fluids. **33**, 033311 (2021). https://doi.org/10.1063/5.0039559
205. Vuorinen, V., Aarnio, M., Alava, M., et al.: Modelling aerosol transport and virus exposure with numerical simulations in relation to SARS-CoV-2 transmission by inhalation indoors. Saf. Sci. **130**, 104866 (2020). https://doi.org/10.1016/j.ssci.2020.104866
206. Lei, H., Li, Y., Xiao, S., et al.: Routes of transmission of influenza A H1N1, SARS CoV, and norovirus in air cabin: comparative analyses. Indoor Air. **28**, 394–403 (2018). https://doi.org/10.1111/ina.12445

207. Stephens, B., Azimi, P., Thoemmes, M.S., et al.: Microbial exchange via fomites and implications for human health. Curr. Pollut. Rep. **5**, 198–213 (2019). https://doi.org/10.1007/s40726-019-00123-6
208. Li, S., Eisenberg, J.N., Spicknall, I.H., Koopman, J.S.: Dynamics and control of infections transmitted from person to person through the environment. Am. J. Epidemiol. **170**, 257–265 (2009)
209. Kraay, A.N., Hayashi, M.A., Hernandez-Ceron, N., et al.: Fomite-mediated transmission as a sufficient pathway: a comparative analysis across three viral pathogens. BMC Infect. Dis. **18**, 1–13 (2018)
210. Hu, L., Zhang, J., Xiang, Y., Wang, W.: Neural networks-based aerodynamic data modeling: a comprehensive review. IEEE Access. **8**, 90805–90823 (2020)
211. Masoumi-Verki, S., Haghighat, F., Eicker, U.: A review of advances towards efficient reduced-order models (ROM) for predicting urban airflow and pollutant dispersion. Build. Environ. **108966**, 108966 (2022)
212. Pant, P., Doshi, R., Bahl, P., Barati Farimani, A.: Deep learning for reduced order modelling and efficient temporal evolution of fluid simulations. Phys. Fluids. **33**, 107101 (2021)
213. Calzolari, G., Liu, W.: Deep learning to replace, improve, or aid CFD analysis in built environment applications: a review. Build. Environ. **206**, 108315 (2021)
214. Ivagnes, A., Demo, N., Rozza, G.: Towards a machine learning pipeline in reduced order modelling for inverse problems: neural networks for boundary parametrization, dimensionality reduction and solution manifold approximation. J. Sci. Comput. **95**, 1–24 (2023)
215. Lui, H.F., Wolf, W.R.: Construction of reduced-order models for fluid flows using deep feedforward neural networks. J. Fluid Mech. **872**, 963–994 (2019)
216. Fukami, K., Hasegawa, K., Nakamura, T., et al.: Model order reduction with neural networks: application to laminar and turbulent flows. SN Comput. Sci. **2**, 1–16 (2021)
217. Xiang, S., Fu, X., Zhou, J., et al.: Non-intrusive reduced order model of urban airflow with dynamic boundary conditions. Build. Environ. **187**, 107397 (2021)
218. Chen, W., Wang, Q., Hesthaven, J.S., Zhang, C.: Physics-informed machine learning for reduced-order modeling of nonlinear problems. J. Comput. Phys. **446**, 110666 (2021)
219. Daniel, T., Casenave, F., Akkari, N., Ryckelynck, D.: Model order reduction assisted by deep neural networks (ROM-net). Adv. Model. Simul. Eng. Sci. **7**, 1–27 (2020)
220. Torregrosa, A., Garcıa-Cuevas, L., Quintero, P., Cremades, A.: On the application of artificial neural network for the development of a nonlinear aeroelastic model. Aerosp. Sci. Technol. **115**, 106845 (2021)
221. Li, T., Wu, T., Liu, Z.: Nonlinear unsteady bridge aerodynamics: reduced-order modeling based on deep LSTM networks. J. Wind Eng. Ind. Aerodyn. **198**, 104116 (2020)
222. Quilodrán-Casas, C., Arcucci, R., Pain, C., Guo, Y.: Adversarially trained LSTMs on reduced order models of urban air pollution simulations. arXiv preprint arXiv:210101568 (2021)
223. Quilodrán-Casas, C., Silva, V.L., Arcucci, R., et al.: Digital twins based on bidirectional LSTM and GAN for modelling the COVID-19 pandemic. Neurocomputing. **470**, 11–28 (2022)
224. Nguyen, L., Raissi, M., Seshaiyer, P.: Modeling, analysis and physics informed neural network approaches for studying the dynamics of COVID-19 involving human-human and human-pathogen interaction. Comput. Math. Biophys. **10**, 1–17 (2022)
225. Nguyen, L., Ghosal, A., Nagalia, R., Seshaiyer, P.: Mathematical modelling and simulation using an efficient PINNs algorithm to understand spread of infection in enclosed spaces (2021)
226. Majumdar, R., Karande, S., Vig, L.: DeepEpiSolver: unravelling inverse problems in Covid, HIV, ebola and disease transmission. arXiv preprint arXiv:230314194 (2023)
227. Shaier, S., Raissi, M., Seshaiyer, P.: Data-driven approaches for predicting spread of infectious diseases through DINNs: disease informed neural networks. arXiv preprint arXiv:211005445 (2021)

Real-Time, Systematic Disease Detection on Cruise Ships: Feasibility Assessment for Outbreak Prevention

Bérengère Lebental

1 Introduction

The heavy toll of COVID-19 on the cruise industry has been documented since 2020. Highly publicized outbreaks on cruise ships in the early days of the pandemic were followed by documented cases of the virus being widely spread in their home countries by passengers returning from cruise ships. Cruises were eventually banned in some countries, while on the other hand, increasing travel restrictions reduced the number of cruise passengers from 30 million in 2019 to 6 million in 2020. It is estimated that cruise lines lost $77 billion between March and September 2020 alone [34, 51].

Since then, the cruise industry has recovered, but the pandemics has dramatically highlighted the long-standing issue of disease and epidemic management on ships [30]. A general distinction can be made between respiratory and gastrointestinal diseases [19]. As of May 2023, the American Centers for Disease Control and Prevention list influenza, COVID-19, and the common cold as main respiratory illnesses and Norovirus as the most common digestive illness. In addition, Legionnaires' disease remains a concern with frequent outbreaks.

To ensure health on board, members of the Cruise Lines International Association (CLIA) (95% of the world's ocean-going cruise capacity) claim to follow the guidelines proposed by the American College of Emergency Physicians (ACEP) for cruise ship medical facilities, established in 1995 and regularly revised since then.[1] The World Health Organization also publishes an International Medical Guide

[1] https://www.acep.org/patient-care/policy-statements/health-care-guidelines-for-cruise-ship-medical-facilities, last accessed May 15th, 2023.

B. Lebental (✉)
Université Gustave Eiffel, COSYS, IMSE, Marne-La-Vallée, France
e-mail: berengere.lebental@univ-eiffel.fr

S. Th. Rassia (ed.), *The Blue Book*, https://doi.org/10.1007/978-3-031-48831-3_9

for Ships, which includes the contents of the ship's medicine chest [56]. During the COVID-19 crisis, the European Medicines Agency published (and revised several times) recommendations for managing COVID on board cruise ships [11], which were different from the available recommendations to manage Influenza-like illness.[2]

To prevent outbreaks on board, the obvious question for the ship's physician is to accurately identify a disease carried by passengers or crew. Legionnaires' disease, influenza, or COVID-19 may all present with similar respiratory symptoms, but the recommendations in the event of an outbreak are very different. Time is also a critical factor. Even in the case of a person with a readily identifiable illness, the latency between the onset of symptoms and the visit to the doctor will result in additional contaminations on the ship. A third aspect is exhaustivity of the detection of sick passengers: in the case of a person with mild symptoms, they may not even visit the doctor, though they may be contagious.

To address these three challenges, it is obvious that the whole extent of analytical equipment regularly used ashore for precision diagnostics is not available on board. While the appropriate biological samples can be sent to shore for accurate laboratory analysis, shipping and processing delays mean that the resulting information is most often used for post-crisis management (e.g., insurance claims). On the other hand, the ACEP healthcare guidelines for cruise ships recommend availability of a set of lab testing capabilities (including blood count, urine analysis, several cardiac enzymes). However, the only listed infectious diseases are Malaria, Legionella, Influenza A and B, and HIV, and their lab testing procedures take days. The only rapid screening tests included in healthcare recommendations are presently the ones for COVID-19.

Overall, optimal prevention of outbreaks on cruise ships would be achieved if each individual on board could be accurately diagnosed at the onset of symptoms, with or without a visit to the ship's health center. This chapter discusses whether the current state of the art of point-of-care diagnostics, biosensors, and wearable devices can meet this challenge. A first section discusses point-of-care diagnosis options at the ship's health center. A second section discusses whether diseases could be diagnosed in real time throughout the ship. Two approaches are considered: direct monitoring of individuals or monitoring of their cabins. Finally, the most promising options for disease detection on cruise ships are summarized.

2 Point-of-Care Diagnosis at the Health Center

The most natural place to identify illness on a cruise ship is the ship's health center. In fact, passengers or crew members who feel ill are expected to go directly to the

[2] https://www.cdc.gov/quarantine/cruise/management/guidance-cruise-ships-influenza-updated.html, last consulted May 15th, 2023.

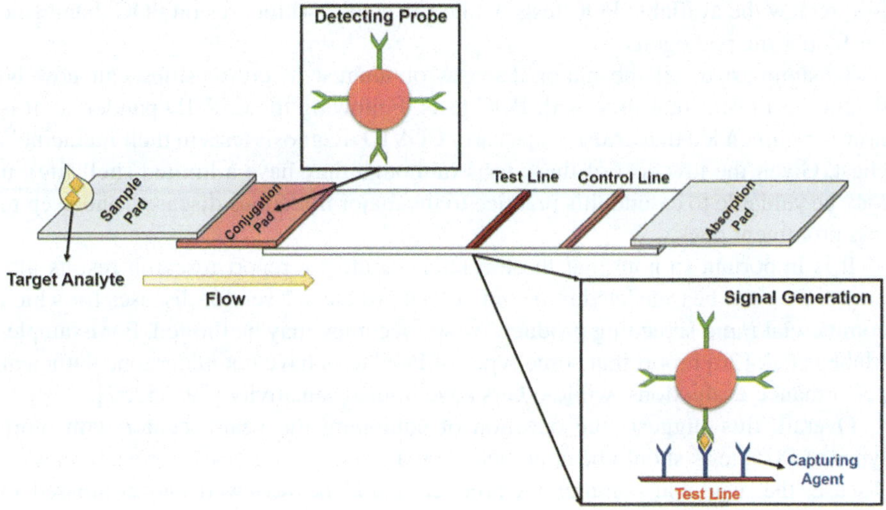

Fig. 1 Schematic representation of lateral flow assay. The target analyte is labeled by the detecting probe, inside the conjugation pad [35]

ship's physician. In addition, crew members can identify passengers with worrisome symptoms and refer them to the doctor as well. The ability to accurately and rapidly diagnose illness in the ship's health center is therefore critical.

Citing [26], point-of-care (POC) tests are "diagnostic tests performed in the clinic, with results available within a short time so that patients can be treated without a return visit." While POC tests are of particular interest in the developing world to detect disease such as malaria and syphilis, they perfectly answer the needs of cruise ships: no need for lab equipment, reduced space, rapid results, low cost. The general principle is to allow a direct color identification of positive results. The relevant sample, such as saliva, urine, or blood, is placed on the test (either dipstick or lateral-flow format) and the test changes color according to the presence of the disease. Most tests rely on immunochromatography, e.g., detection of antigens or antibodies. Figure 1 provides an illustration of a POC test in the lateral flow format.

Naseri et al. [35] provides examples of POC tests for virus detection, with particular emphasis on solutions targeting COVID-19. Solutions to identify the different influenza viruses are also described. Regarding gastrointestinal diseases, one of the main concerns on board cruise ships is to detect norovirus. POC diagnostics are available for this as well, as detailed in Chand and Neethirajan [6], Kim et al. [21], Qian et al. [41], and Weng and Neethirajan [54].

There is also a clear need to rapidly track diseases caused by environmental factors: legionella in HVAC or water systems, *Escherichia coli*, Salmonella or Listeriosis in food or water. In this case, the identification of the disease is needed to search for its root cause on the ship. Chen et al. [9], Roy et al. [45], and Sohrabi et al.

[48] review the available POC tests, while Liu et al. [24] focuses on POC diagnostic for foodborne pathogens.

To summarize, all the major diseases of interest to cruise ships can now be diagnosed in near real-time with POC tests. Following the COVID pandemic, it is now recommended that cruise ships carry COVID diagnosis tests in their medicine's chest. Given the low cost of these tests (although they have a limited shelf life), it may be valuable to extend this practice to the major infectious diseases, and even to less prominent ones.

It is important to note that these references mostly report research results that may or may not be translated into commercial products. Even for diseases for which commercial rapid screening products exists, accuracy may be limited. For example, Mabey et al. [26] report that some types of POC tests have not undergone sufficient performance evaluations, while others have limited sensitivity [39, 55, 57].

Overall, this suggests the question of equipping the health center with more types of POC tests should be approached on a case-by-case basis; for each relevant disease, the available commercial products could be reviewed and compared in terms of accuracy, response time, price, and shelf life. Based on this, an appropriate strategy could be developed; number of tests to have on board; how to proceed (from an epidemic mitigation point of view) in the case of both positive and negative tests.

3 Real-Time, Systematic and Selective Disease Identification

The main limitation of point-of-care diagnosis at the health center is that it requires decisive action by the individual to go to the health center, which has financial and time implications. This action by the individual is usually triggered only in the case of severe symptoms, long after they have become contagious or long after the environmental cause has started. Thus, by its very nature, diagnosis at the health center cannot provide a complete or early identification of individuals with a specific disease.

For early and systematic disease detection, the ideal solution is to continuously monitor the presence/absence of one or more selected diseases at the individual passenger or crew level in real time. As a starting point, we note [2], in another chapter of the same book, detailed possible strategies for real-time symptom detection throughout the ship. This is a valuable first step in epidemic management. The present section goes beyond symptom detection and discuss whether disease can be detected selectively and systematically throughout the ship. There are two ways to do this on board a cruise ship, either by monitoring individuals directly with a wearable health monitor, or by monitoring their cabin. This section discusses the feasibility of the same in the context of a cruise ship.

3.1 Smart Wearable Devices

The rapid development of smart wearable devices (SWDs) over the past 15 years has strongly attracted the attention of the healthcare community as a means to provide noninvasive, continuous health monitoring at low cost [7, 32]. While SWDs encompass a wide range of very different solutions (from smart tattoos to smart helmets), their common denominator is the inclusion of one or more sensors that are packaged in a solution that can be worn on the body. Healthcare SWDs can be broadly divided into two groups: those that monitor symptoms and general physiological information such as heartbeat or temperature, and those that monitor one or more specific biomarkers of a disease. The first group generally includes solutions that are either commercially available or for which all the building blocks (sensors, electronics . . .) are on the market (i.e., they could theoretically be on the market in the near future). The second group tends to include solutions for which one or more of the key components is still a major research challenge.

As described in detail in Chan et al. [7] and Mukherjee et al. [32], a major area of application in healthcare has been chronic diseases. Until the pandemic, the use of SWDs to detect infectious disease was rare. COVID-19 brought to light a body of previous work on the topic, while motivating rapid new developments, as reviewed by Mohammadzadeh et al. [31]: a watch-like device for COVID detection using body surface temperature and heart rate was reported [10], while a Fitbit wearable device was proposed for influenza detection [42]; various wearable solutions were also reported for Ebola (3), Chikungunya (1), and Dengue (1); the review also includes SWD for general "infectious diseases". Since this review in 2020, [33] also described a prototype solution for COVID screening including heartbeat, O2 saturation, temperature, and cough sensors. Al-Halhouli et al. [1] extended the review, with a specific focus on SWDs based on printed electronics.

Overall, the studies suggest the feasibility of disease identification through symptom monitoring by SWD. One limitation is that the studies are rarely based on already commercial SWDs – which would be immediately deployable in a cruise ship context. As an exception, the study by Radin et al. [42] is particularly interesting because it relied on an already commercial SWD that is widely used in the USA and collected data from a significant number of users (47000+). The data from the SWDs (sleep duration and resting heart rate) showed a very good correlation across the USA with the "Influenza-like-Illness" evolution curves by state published by the CDC (Fig. 2). This suggests a quite mature solution. On the other hand, by focusing on aggregated "Influenza-like-Illness" data, the study does not claim selectivity between different types of influenza-like illness.

Selectivity in disease detection based on symptoms and general physiological monitoring is a critical bottleneck in the literature on SWDs. As reviewed by Al-Halhouli et al. [1], the symptoms and vital signs monitored by physicians to diagnose a number of emerging viral diseases (COVID-19, Crimean-Congo Hemorrhagic Fever, Ebola Virus Disease, Marburg Virus Disease, Lassa Hemorrhagic Fever, Middle East Respiratory Syndrome, Severe Acute Respiratory Syndrome,

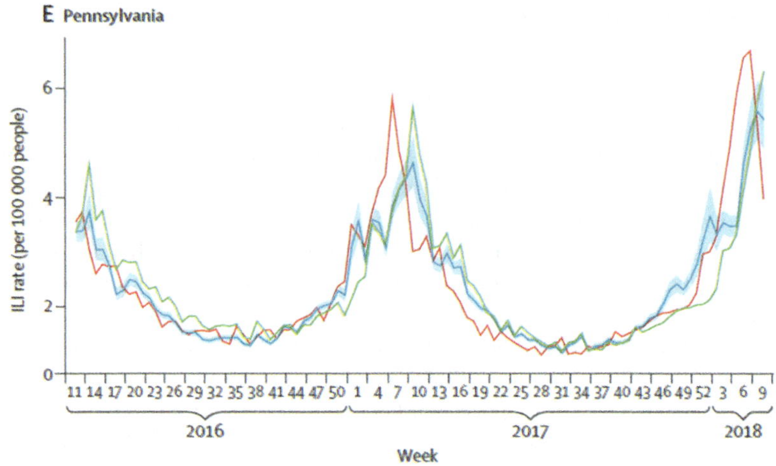

Fig. 2 Weekly CDC Influenza-like illness (ILI) rates (red), predicted ILI rates from the baseline model H0 model (green) and for an optimized model H1 (blue, with 95% confidence intervals in light blue), for Pennsylvania. Data are from March 16, 2016, to March 1, 2018. CDC Centers for Disease Control and Prevention, ILI influenza-like illness. (Reproduced from Radin et al. [42])

Nipah Virus disease, and Rift Valley fever) differ between diseases. And most of these symptoms and vital signs can be monitored by existing commercial or pre-commercial SWDs. Thus, in theory, selectivity could be achieved. However, this has not yet been proved in the general case.

In the specific case of cough-inducing diseases (including COVID-19), this challenge of selectivity has been successfully addressed based on sound monitoring. Machine-learning-based processing of respiratory sounds (cough, breath, voice) allows differentiation between respiratory diseases [16]. This approach has shown positive results in field trials with dozens of patients for infectious (bronchitis and COVID-19) and non-infectious (asthma) respiratory diseases [3, 36]. Considering sound sensors are available at low cost, this type of strategy has a great potential for generalization. At this time, proof of concepts has been already successful with smart phones and smart watches [16] (Fig. 3).

To ensure selectivity, the most common strategy is to incorporate sensors into the SWDs sensors that will monitor biomarkers of the disease. This requires sampling of bodily fluids, such as blood, saliva or sweat, or breath analysis. Monitoring either sweat or breath is the most promising strategy in the context of cruise ships, as it is minimally intrusive (compared to blood or saliva sampling).

For sweat monitoring [53], the principle is to stick smart tape on the skin. Smart tattoos are also an option. Either type of devices collects sweat and analyzes it for specific biomarkers. To collect adequate amounts of sweat, additional chemicals can be used to increase sweat production. So far, the main application is the monitoring of chronic diseases [28]: diabetes, cancer, cystic fibrosis, etc. Applications for infectious disease detection are very rare. Recently, Jagannath et al. [17, 18]

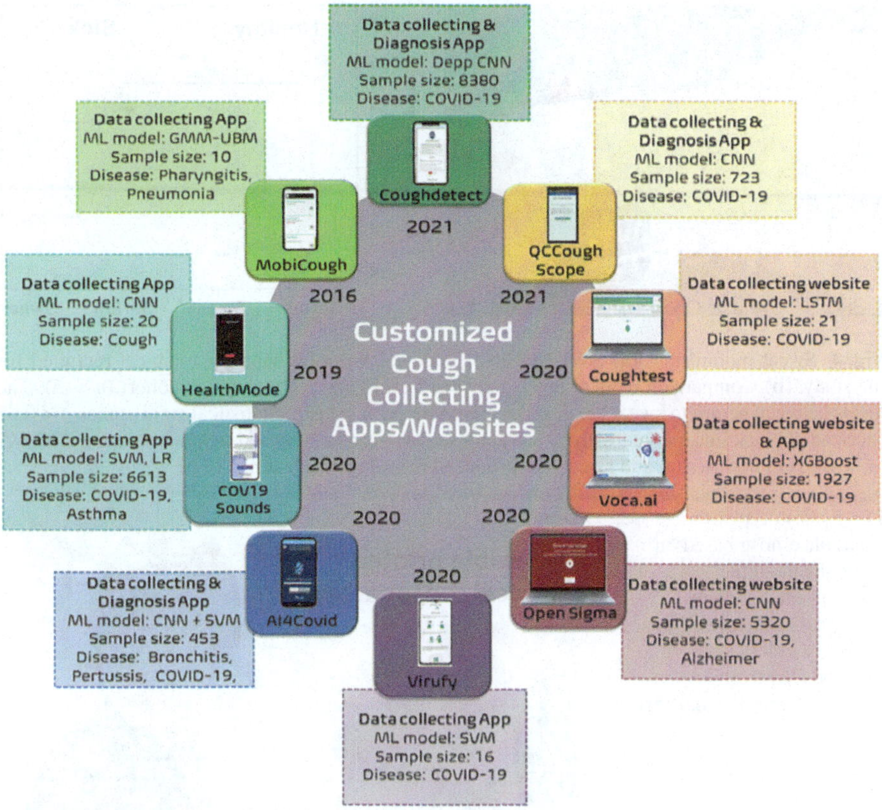

Fig. 3 Customized cough detection and disease diagnosing applications. (Reproduced from Ijaz et al. [16])

demonstrated the possibility of tracking inflammatory markers of disease in sweat using on-body sensing with a watch-like device, a solution relevant to COVID or influenza monitoring (Fig. 4). This study is a research paper, so significant translational research is needed before it can be used and validated under field conditions However, it is important because it opens up the way to accurate (by design because of biomarker detection) and exhaustive (because smart tape could be very cheap) tracking of a specific virus in large numbers of people.

Another promising approach is breath monitoring. Hu and Koo [15] reported on a face mask for COVID detection that changes color in response to COVID exposure (from one's own breath or from someone else's). While this is not directly a smart solution, it is a solution that could have value on a cruise ship.

Both sweat and breath carry odors that can be quantified by measuring their concentration in volatile organic compounds (VOCs). It has been demonstrated under clinical conditions and with laboratory analytical equipment that the VOC composition of breath and sweat carries selective information about some infectious

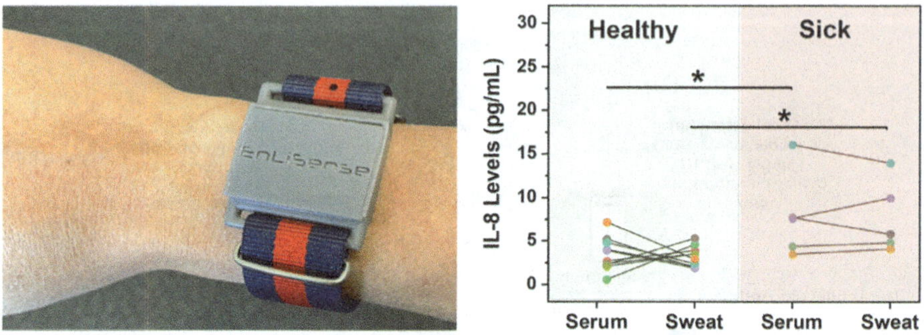

Fig. 4 Sweat monitoring: (**a**) SWEATSENSER device worn on hand by a subject recruited for the study. (**b**) Comparison of IL-8 levels between healthy ($n = 10$) and sick cohort ($n = 5$). The levels between serum to sweat are also compared for each subject. Statistical significance $p < 0.05$ (horizontal black line marked with *) was achieved between healthy and sick cohort using ANOVA ($\alpha = 0.05$) in both serum and sweat. (Reproduced from Jagannath et al. [17])

Fig. 5 Prototype of a wearable e-nose based on using the flexible inkjet-printed chemical sensor array integrated in ZigBee wireless network. (Reproduced from Lorwongtragool et al. [25])

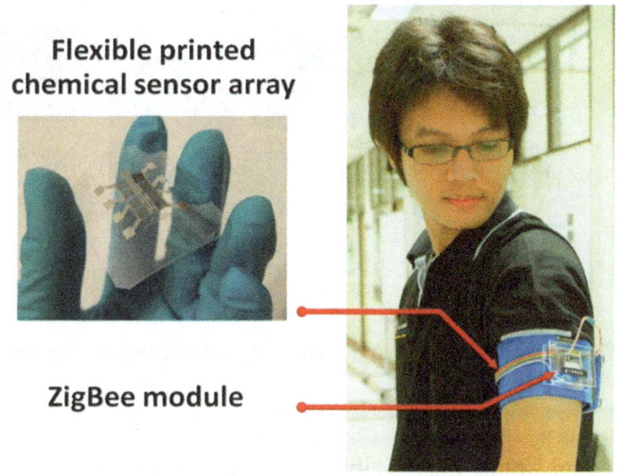

diseases [13, 47]. Based on this principle, Malikhah et al. [27] demonstrated the feasibility of detecting COVID by measuring the odor of axillary sweat samples using an electronic nose (e-nose) for volatile organic compounds. The main challenge of this approach is the technological maturity of e-noses for VOC monitoring, especially in the context of SWDs. It is still a major topic of research, as is reviewed by Cheng et al. [8]. While studies by Lorwongtragool et al. [25] (Fig. 5) and Seesaard et al. [46] provide examples of e-nose applications for health monitoring, there does not yet appear to be any commercial or pre-commercial solution available that is suitable for disease tracking.

3.2 Cabin Monitoring

3.2.1 Correlation of Symptoms

Alternatively, instead of continuously monitoring a person with an SWD, one may choose to monitor their cabin. Each person on a cruise ship will spend time in their cabin on a regular basis, and providing sensor solutions for the cabin is less challenging (in terms of size and ruggedness requirements) than for a smart wearable device. Some of the monitoring options discussed with SWDs can be directly applied to the cabin as well, while the cabin context offers additional monitoring opportunities.

On the topic of cough monitoring, which is very promising for selective disease recognition with SWDs, the approach also works with in situ, fixed sound sensors in the cabin [16]. It may even have advantages over the use of SWDs: reduced ambient noise, larger size allowing higher quality of sensors, possibility to track additional symptoms through noise (e.g., increased frequency of toilet flushing).

On the other hand, odor monitoring for disease identification, which is not yet feasible with SWDs, is not yet mature in the cabin context either. In fact, highly sensitive, compact VOC sensors for indoor use are needed. While there are many ongoing studies, there are no market-ready solution yet [20, 50]. The first pre-commercial solutions are likely to be more suitable for point-of-care testing [4] than for continuous environmental monitoring of disease-related odors.

3.2.2 Direct Disease Detection

All of the methods described above, including all but two of the solutions with SWDs (the face mask that changes color in the presence of SARS-COV-2 [15]; the sweat monitor of Jagannath et al. [17, 18]), are based on the indirect detection of a disease by correlating symptoms and vital signs. In the present section, the feasibility of direct detection, in the cabin, of the pathogen itself or of selective biomarkers (RNA, DNA, antibodies . . .) will be discussed.

Let us first discuss the indoor localization of pathogens: for airborne and droplet-borne pathogens they have been detected on surfaces or in the air; for pathogens with biomarkers or direct presence in saliva, sweat, urine, and feces, they have been detected in bathroom effluents (gray water from sinks and showers; black water from toilets) [40]. The ability to detect SARS-COV-2 in air sampled directly in a room [43] or in the room HVAC system [49] was demonstrated (Fig. 6), while J. Pan et al. [37] identified the room surfaces where SARS-COV-2 is most present: air filters, sink floor, and countertop. The presence of SARS-COV-2 in wastewater has been well documented since the seminal work of Randazzo et al. [44].

Indoor pathogen detection is composed of three steps: sampling, sample preparation, and detection. The sample preparation stage – whenever required – include transfer of the sample in a medium appropriate for measurement and/or preconcentration.

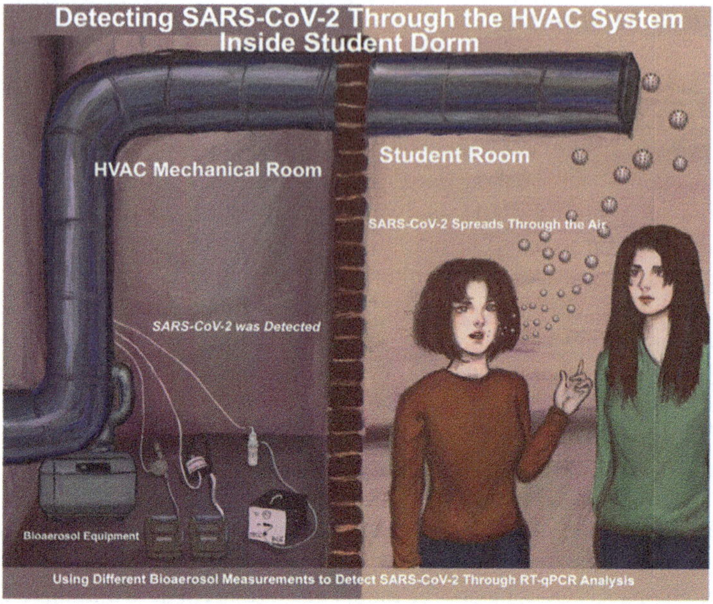

Fig. 6 Illustration of HVAC-system based detection of SARS-COV-2 [49]

Regarding sampling, automated air sampling has been available for some time; the numerous options are described in detail in Krokhine et al. [22] and M. Pan et al. [38], for example (Fig. 7). Air filters are also natural collectors of air-borne pathogens. Automated urine and stool sampling in the toilet has recently emerged and is now at prototype level [14, 29, 52] (Fig. 8). Let us remark here that, in the context of a cruise ship, these sampling solutions can be applied directly at cabin level, or by groups of cabins (e.g., connected to the same HVAC or blackwater system), or for the full ship.

On the detection side, a very large number of reviews cover the range of biosensors for identification of infectious diseases, such as Pilevar et al. [40] for water-borne pathogens, Bhardwaj et al. [5], Fronczek and Yoon [12], and Pilevar et al. [40] for airborne diseases (Fig. 9). Limits of detection are now extremely low, e.g., single-virus-detection of SARS-COV-2 is possible with a portable solution based on carbon nanotube sensors [23] (Fig. 10).

So, in principle, there is no scientific barrier to achieving automated sampling and then pathogen detection. However, in practice, solutions that combine an automated sampling system with a biosensor solution are extremely rare. The automated urine analysis solution developed by the Yellosis[3] company (part of the Samsung startup ecosystem) was the only example found as of May 2023 (and it does not even target infectious diseases). Most likely, various companies are developing their

[3] https://cym702.com/app.php, last accessed May 15th, 2023.

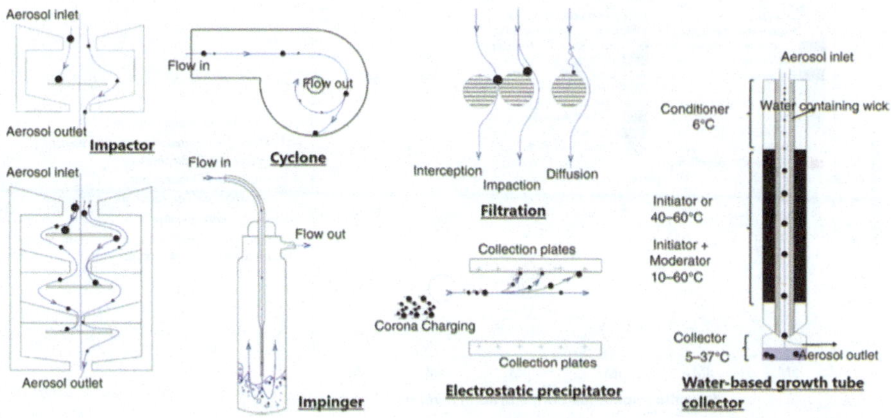

Fig. 7 Solutions for collection of air samples. (Reproduced from Pan et al. [38])

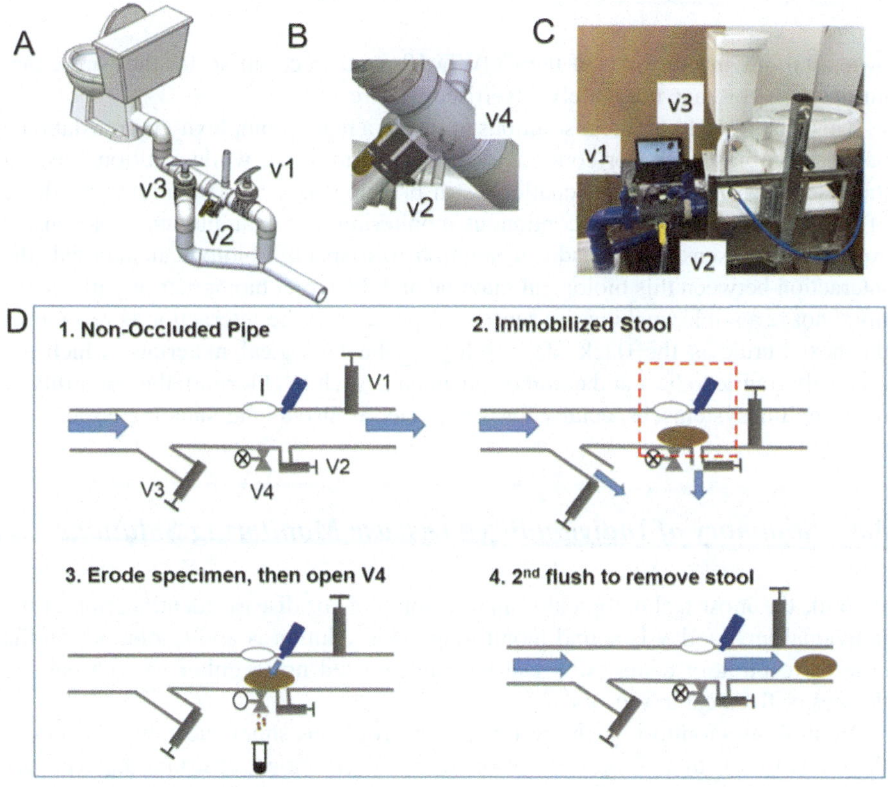

Fig. 8 Schematics (**a**, **b**), prototype image (**c**) and principles of operation (**d**) of an automated, toilet-integrated stool sample collection system. (Reproduced from Grego et al. [14])

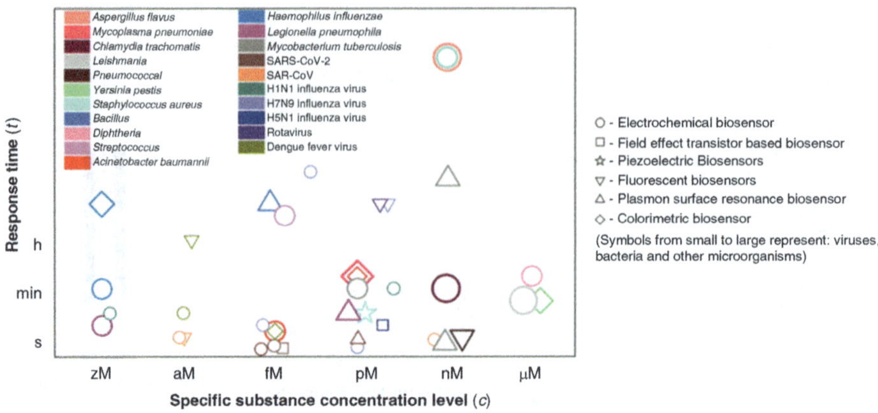

Fig. 9 Classification of biosensors for infectious disease detection, by transduction type, limit of detection and response time. (Reproduced from Pilevar et al. [40])

own solutions in the wake of the COVID-19 pandemic, but so far they are either unsuccessful or have not widely advertised their results.

This lack of technological solutions points to a major complexity in translational research on biosensors for continuous disease monitoring: while solution exists at the research level, there are challenges in bringing biosensor solutions to a level of maturity compatible with continuous monitoring. For example, most biosensors couple a very sensitive transducer solution to a specific biological material; the interaction between this biological material and the target biomarker or pathogen is often not reversible, making the sensors disposable; if the interaction is reversible, the next hurdle is the (lack of) stability of the biological materials, which are relatively fragile to heat and common chemicals (such as chlorine); the durability of the sampling systems, especially for wastewater, is also a long-standing issue.

3.3 Summary of Individualized Disease Monitoring Solutions

Overall, the most technologically mature solution for disease identification at the individual level today is sound monitoring. This solution is applicable only in the case of respiratory diseases. It can be implemented using either smart wearable devices or fixed cabin sensors.

Vital signs monitoring via commercially available smart wearable devices is the second most mature option, but there are uncertainties regarding the accuracy of disease identification. While detecting whether an individual has a disease is straightforward, the capability to distinguish between diseases remains unclear and requires further investigation.

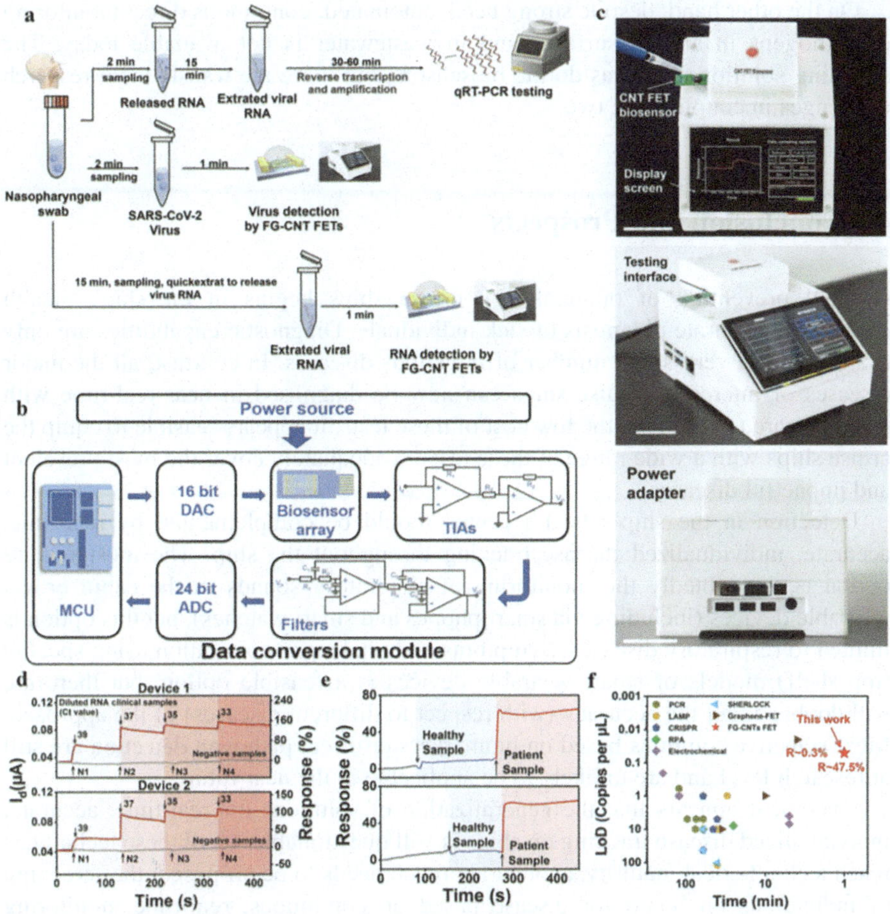

Fig. 10 Real-time detection of SARS-CoV-2 antigen and nucleic acid. (**a**) Comparison of the workflow for detecting COVID-19 by qRT-PCR and multifunctionalized floating gate carbon nanotube field effect transistor (FG-CNT FET) FG-CNT FETs. (**b**) Data conversion module of the COVID-19 testing system. (**c**) Prototype demonstration of the COVID-19 testing system, including test modules and power adapter sections (the scale bar: up: 4 cm, middle: 3.5 cm; bottom: 5 cm). (**d**) Drain current versus time curve of diluted positive samples with concentrations from 0.05 copies/μL (Ct value is 39–100%) positive samples. (**e**) Dynamic response signal detection in healthy and patient samples in UTM. (**f**) Comparison of FG-CNT FETs (this work) with other SARS-CoV-2 RNA detection (LAMP: loop-mediated isothermal amplification; CRISPR: clustered regularly interspaced short palindromic repeats; RPA: recombinase polymerase amplification; SHERLOCK: specific high-sensitivity enzymatic reporter unlocking. R: response). (Reproduced from Liang et al. [23])

At prototype level, selective detection of an infectious disease by monitoring selected biomarkers in sweat has proven feasible, although the time-to-market of such a solution remains to be evaluated.

On the other hand, despite strong need, automated, continuous direct monitoring of pathogens in air, on surfaces and in wastewater is not available today. The sampling solutions exist, as do the biosensors, but there are translational research challenges in coupling the two.

4 Conclusion and Prospects

Optimal prevention of outbreaks on cruise ships begins in the ship's health center with accurate diagnosis of sick individuals. Diagnostic capabilities are only available for a very small number of infectious diseases. In contrast, all the major diseases of interest to cruise ships can now be diagnosed in near real-time with point-of-care tests. Given the low cost of these tests, it appears feasible to equip the cruise ships with a wide range of these tests to adequately cover the most prevalent and impactful diseases.

Detection in the ship's health center should be complemented by real-time, accurate, individualized disease tracking throughout the ship. The most mature option is undoubtedly the monitoring of respiratory sounds in the cabin or via wearable devices (including via smart phones and smart watches), but this option is limited to respiratory diseases. Symptoms and vital signs correlation using specific (top shelf) models of smart wearable devices is a feasible option, but there are still doubts about the accuracy (with respect to different diseases) of the approach. More selective solutions based on biomarkers or direct pathogen detection are still at research level and are unlikely to be applicable in the near future.

Overall, it appears that the generalization of solutions for real-time, accurate, individualized disease tracking on ships is still questionable. Until these techniques reach technological maturity, another approach needs to be proposed: the screening of individuals on board for disease based on continuous, real-time monitoring of symptoms. As is described in Bruno Almeida et al. [2] this can be achieved with current commercial technologies, either with smart wearable devices, with fixed cabin sensors and/or with sensors deployed throughout the ship. A positively screened individual would be offered an immediate visit to the health center. There, point-of-care diagnosis would be performed straightaway to assess the risk of the symptoms being caused by an infectious disease. This solution is not perfect, as it requires action on the part of the screened individual; however, proper configuration of the symptom-based screening algorithms would ensure early detection of sick passengers throughout the whole ship, which in turn would allow for early activation of epidemic mitigation measures.

Acknowledgments This chapter acknowledges funding from European Union for the project HS4U under GA number 101069937.

References

1. Al-Halhouli, A., Albagdady, A., Alawadi, J., Abeeleh, M.A.: Monitoring symptoms of infectious diseases: perspectives for printed wearable sensors. Micromachines. **12**(6), 620 (2021). https://doi.org/10.3390/mi12060620
2. Almeida, B., Mateus, M., Pacheco, M., Malo, P., Teixeira, T.: Technology approaches for cruise ship disease propagation monitoring. In: TBD. Springer Book (n.d.)
3. Andreu-Perez, J., Perez-Espinosa, H., Timonet, E., Kiani, M., Giron-Perez, M.I., Benitez-Trinidad, A.B., Jarchi, D., Rosales-Perez, A., Gatzoulis, N., Reyes-Galaviz, O.F., Torres-Garcia, A., Reyes-Garcia, C.A., Ali, Z., Rivas, F.: A generic deep learning based cough analysis system from clinically validated samples for point-of-need Covid-19 test and severity levels. IEEE Trans. Serv. Comput. **15**(3), 1220–1232 (2022). https://doi.org/10.1109/TSC.2021.3061402
4. Beduk, T., Durmus, C., Hanoglu, S.B., Beduk, D., Salama, K.N., Goksel, T., Turhan, K., Timur, S.: Breath as the mirror of our body is the answer really blowing in the wind? Recent technologies in exhaled breath analysis systems as non-invasive sensing platforms. TrAC Trends Anal. Chem. **143**, 116329 (2021). https://doi.org/10.1016/j.trac.2021.116329
5. Bhardwaj, S.K., Bhardwaj, N., Kumar, V., Bhatt, D., Azzouz, A., Bhaumik, J., Kim, K.-H., Deep, A.: Recent progress in nanomaterial-based sensing of airborne viral and bacterial pathogens. Environ. Int. **146**, 106183 (2021). https://doi.org/10.1016/j.envint.2020.106183
6. Chand, R., Neethirajan, S.: Microfluidic platform integrated with graphene-gold nano-composite aptasensor for one-step detection of norovirus. Biosens. Bioelectron. **98**, 47–53 (2017). https://doi.org/10.1016/j.bios.2017.06.026
7. Chan, M., Estève, D., Fourniols, J.-Y., Escriba, C., Campo, E.: Smart wearable systems: current status and future challenges. Artif. Intell. Med. **56**(3), 137–156 (2012). https://doi.org/10.1016/j.artmed.2012.09.003
8. Cheng, L., Meng, Q.-H., Lilienthal, A.J., Qi, P.-F.: Development of compact electronic noses: a review. Meas. Sci. Technol. **32**(6), 062002 (2021). https://doi.org/10.1088/1361-6501/abef3b
9. Chen, X.-F., Zhao, X., Yang, Z.: Aptasensors for the detection of infectious pathogens: design strategies and point-of-care testing. Microchim. Acta. **189**(12), 443 (2022). https://doi.org/10.1007/s00604-022-05533-w
10. Chung, Y.-T., Yeh, C.-Y., Shu, Y.-C., Chuang, K.-T., Chen, C.-C., Kao, H.-Y., Ko, W.-C., Chen, P.-L., Ko, N.-Y.: Continuous temperature monitoring by a wearable device for early detection of febrile events in the SARS-CoV-2 outbreak in Taiwan, 2020. J. Microbiol. Immunol. Infect. **53**(3), 503–504 (2020). https://doi.org/10.1016/j.jmii.2020.04.005
11. European Medicines Agency. (2022). Advice for cruise ship operators for preparedness and response to an outbreak of COVID-19. https://www.healthygateways.eu/Portals/0/plcdocs/EU_HEALTHY_GATEWAYS_COVID-19_MARITIME_CRUISE_30_4_2021.pdf?ver=2021-04-30-092015-073
12. Fronczek, C.F., Yoon, J.-Y.: Biosensors for monitoring airborne pathogens. SLAS Technol. **20**(4), 390–410 (2015). https://doi.org/10.1177/2211068215580935
13. Ghosh, C., Leon, A., Koshy, S., Aloum, O., Al-Jabawi, Y., Ismail, N., Weiss, Z.F., Koo, S.: Breath-based diagnosis of infectious diseases. Clin. Lab. Med. **41**(2), 185–202 (2021). https://doi.org/10.1016/j.cll.2021.03.002
14. Grego, S., Welling, C.M., Miller, G.H., Coggan, P.F., Sellgren, K.L., Hawkins, B.T., Ginsburg, G.S., Ruiz, J.R., Fisher, D.A., Stoner, B.R.: A hands-free stool sampling system for monitoring intestinal health and disease. Sci. Rep. **12**(1), 10859 (2022). https://doi.org/10.1038/s41598-022-14803-9
15. Hu, B., Koo, S.: Development of wearable device designs for respiratory infection prevention. Int. J. Cloth. Sci. Technol. **34**(5), 764–781 (2022). https://doi.org/10.1108/IJCST-07-2021-0094
16. Ijaz, A., Nabeel, M., Masood, U., Mahmood, T., Hashmi, M.S., Posokhova, I., Rizwan, A., Imran, A.: Towards using cough for respiratory disease diagnosis by leveraging arti-

ficial intelligence: a survey. Inf. Med. Unlock. **29**, 100832 (2022). https://doi.org/10.1016/j.imu.2021.100832
17. Jagannath, B., Lin, K., Pali, M., Sankhala, D., Muthukumar, S., Prasad, S.: Temporal profiling of cytokines in passively expressed sweat for detection of infection using wearable device. Bioeng. Transl. Med. **6**(3), e10220 (2021). https://doi.org/10.1002/btm2.10220
18. Jagannath, B., Pali, M., Lin, K., Sankhala, D., Naraghi, P., Muthukumar, S., Prasad, S.: Novel approach to track the lifecycle of inflammation from chemokine expression to inflammatory proteins in sweat using electrochemical biosensor. Adv. Mater. Technol. **7**(8), 2101356 (2022). https://doi.org/10.1002/admt.202101356
19. Kak, V.: Infections in confined spaces: cruise ships, military barracks, and college dormitories. Infect. Dis. Clin. N. Am. **21**(3), 773–784 (2007). https://doi.org/10.1016/j.idc.2007.06.004
20. Khatib, M., Haick, H.: Sensors for volatile organic compounds. ACS Nano. **16**(5), 7080–7115 (2022). https://doi.org/10.1021/acsnano.1c10827
21. Kim, S., Lee, S., Lee, H.J.: An aptamer-aptamer sandwich assay with nanorod-enhanced surface plasmon resonance for attomolar concentration of norovirus capsid protein. Sensors Actuators B Chem. **273**, 1029–1036 (2018). https://doi.org/10.1016/j.snb.2018.06.108
22. Krokhine, S., Torabi, H., Doostmohammadi, A., Rezai, P.: Conventional and microfluidic methods for airborne virus isolation and detection. Colloids Surf. B: Biointerfaces. **206**, 111962 (2021). https://doi.org/10.1016/j.colsurfb.2021.111962
23. Liang, Y., Xiao, M., Xie, J., Li, J., Zhang, Y., Liu, H., Zhang, Y., He, J., Zhang, G., Wei, N., Peng, L., Ke, Y., Zhang, Z.: Amplification-free detection of SARS-CoV-2 down to single virus level by portable carbon nanotube biosensors. Small. **19** (2023). https://doi.org/10.1002/smll.202208198
24. Liu, S., Zhao, K., Huang, M., Zeng, M., Deng, Y., Li, S., Chen, H., Li, W., Chen, Z.: Research progress on detection techniques for point-of-care testing of foodborne pathogens. Front. Bioeng. Biotechnol. **10** (2022). https://doi.org/10.3389/fbioe.2022.958134
25. Lorwongtragool, P., Sowade, E., Watthanawisuth, N., Baumann, R., Kerdcharoen, T.: A novel wearable electronic nose for healthcare based on flexible printed chemical sensor array. Sensors. **14**(10), 19700–19712 (2014). https://doi.org/10.3390/s141019700
26. Mabey, D., Peeling, R.W., Ustianowski, A., Perkins, M.D.: Diagnostics for the developing world. Nat. Rev. Microbiol. **2**(3), 231–240 (2004). https://doi.org/10.1038/nrmicro841
27. Malikhah, Sarno, R., Inoue, S., Ardani, M.S.H., Purbawa, D.P., Sabilla, S.I., Sungkono, K.R., Fatichah, C., Sunaryono, D., Bakhtiar, A., Libriansyah, Prakoeswa, C.R.S., Tinduh, D., Hernaningsih, Y.: Detection of infectious respiratory disease through sweat from axillary using an E-nose with stacked deep neural network. IEEE Access. **10**, 51285–51298 (2022). https://doi.org/10.1109/ACCESS.2022.3173736
28. Marques-Deak, A., Cizza, G., Eskandari, F., Torvik, S., Christie, I.C., Sternberg, E.M., Phillips, T.M.: Measurement of cytokines in sweat patches and plasma in healthy women: validation in a controlled study. J. Immunol. Methods. **315**(1–2), 99–109 (2006). https://doi.org/10.1016/j.jim.2006.07.011
29. Mccord et al.: Toilet based urine analysis system (2019)
30. Minooee, A., Rickman, L.S., Rickman.: Infectious diseases on cruise ships. Clin. Infect. Dis. **29**(4), 737–743 (1999)
31. Mohammadzadeh, N., Gholamzadeh, M., Saeedi, S., Rezayi, S.: The application of wearable smart sensors for monitoring the vital signs of patients in epidemics: a systematic literature review. J. Ambient. Intell. Humaniz. Comput. **14**, 6027 (2020). https://doi.org/10.1007/s12652-020-02656-x
32. Mukherjee, S., Suleman, S., Pilloton, R., Narang, J., Rani, K.: State of the art in smart portable, wearable, ingestible and implantable devices for health status monitoring and disease management. Sensors. **22**(11), 4228 (2022). https://doi.org/10.3390/s22114228
33. Mukhtar, H., Rubaiee, S., Krichen, M., Alroobaea, R.: An IoT framework for screening of COVID-19 using real-time data from wearable sensors. Int. J. Environ. Res. Public Health. **18**(8), 4022 (2021). https://doi.org/10.3390/ijerph18084022

34. Muritala, B.A., Hernández-Lara, A.-B., Sánchez-Rebull, M.-V., Perera-Lluna, A.: #CoronavirusCruise: impact and implications of the COVID-19 outbreaks on the perception of cruise tourism. Tour. Manag. Perspect. **41**, 100948 (2022). https://doi.org/10.1016/j.tmp.2022.100948
35. Naseri, M., Ziora, Z.M., Simon, G.P., Batchelor, W.: ASSURED-compliant point-of-care diagnostics for the detection of human viral infections. Rev. Med. Virol. **32**(2) (2022). https://doi.org/10.1002/rmv.2263
36. Pal, A., Sankarasubbu, M.: Pay attention to the cough. Proceedings of the 36th Annual ACM Symposium on Applied Computing, 620–628. https://doi.org/10.1145/3412841.3441943 (2021)
37. Pan, J., Hawks, S.A., Prussin, A.J., Duggal, N.K., Marr, L.C.: SARS-CoV-2 on surfaces and HVAC filters in dormitory rooms. Environ. Sci. Technol. Lett. **9**(1), 71–76 (2022). https://doi.org/10.1021/acs.estlett.1c00892
38. Pan, M., Lednicky, J.A., Wu, C.-Y.: Collection, particle sizing and detection of airborne viruses. J. Appl. Microbiol. **127**(6), 1596–1611 (2019). https://doi.org/10.1111/jam.14278
39. Pate, M.S., Dixon, P.B., Hardy, K., Crosby, M., Hook, E.W.: Evaluation of the biostar chlamydia OIA assay with specimens from women attending a sexually transmitted disease clinic. J. Clin. Microbiol. **36**(8), 2183–2186 (1998). https://doi.org/10.1128/JCM.36.8.2183-2186.1998
40. Pilevar, M., Kim, K.T., Lee, W.H.: Recent advances in biosensors for detecting viruses in water and wastewater. J. Hazard. Mater. **410**, 124656 (2021). https://doi.org/10.1016/j.jhazmat.2020.124656
41. Qian, W., Huang, J., Wang, T., Fan, C., Kang, J., Zhang, Q., Li, Y., Chen, S.: Ultrasensitive and visual detection of human norovirus genotype GII.4 or GII.17 using CRISPR-Cas12a assay. Virol. J. **19**(1), 150 (2022). https://doi.org/10.1186/s12985-022-01878-z
42. Radin, J.M., Wineinger, N.E., Topol, E.J., Steinhubl, S.R.: Harnessing wearable device data to improve state-level real-time surveillance of influenza-like illness in the USA: a population-based study. Lancet Digital Health. **2**(2), e85–e93 (2020). https://doi.org/10.1016/S2589-7500(19)30222-5
43. Rahmani, A.R., Leili, M., Azarian, G., Poormohammadi, A.: Sampling and detection of corona viruses in air: a mini review. Sci. Total Environ. **740**, 140207 (2020). https://doi.org/10.1016/j.scitotenv.2020.140207
44. Randazzo, W., Truchado, P., Cuevas-Ferrando, E., Simón, P., Allende, A., Sánchez, G.: SARS-CoV-2 RNA in wastewater anticipated COVID-19 occurrence in a low prevalence area. Water Res. **181**, 115942 (2020). https://doi.org/10.1016/j.watres.2020.115942
45. Roy, S., Arshad, F., Eissa, S., Safavieh, M., Alattas, S.G., Ahmed, M.U., Zourob, M.: Recent developments towards portable point-of-care diagnostic devices for pathogen detection. Sensors Diagn. **1**(1), 87–105 (2022). https://doi.org/10.1039/D1SD00017A
46. Seesaard, T., Seaon, S., Lorwongtragool, P., & Kerdcharoen, T. (2014). On-cloth wearable E-nose for monitoring and discrimination of body odor signature. 2014 IEEE Ninth International Conference on Intelligent Sensors, Sensor Networks and Information Processing (ISSNIP), 1–5. https://doi.org/10.1109/ISSNIP.2014.6827634
47. Sethi, S., Nanda, R., Chakraborty, T.: Clinical application of volatile organic compound analysis for detecting infectious diseases. Clin. Microbiol. Rev. **26**(3), 462–475 (2013). https://doi.org/10.1128/CMR.00020-13
48. Sohrabi, H., Majidi, M.R., Fakhraei, M., Jahanban-Esfahlan, A., Hejazi, M., Oroojalian, F., Baradaran, B., Tohidast, M., de la Guardia, M., Mokhtarzadeh, A.: Lateral flow assays (LFA) for detection of pathogenic bacteria: A small point-of-care platform for diagnosis of human infectious diseases. Talanta. **243**, 123330 (2022). https://doi.org/10.1016/j.talanta.2022.123330
49. Sousan, S., Fan, M., Outlaw, K., Williams, S., Roper, R.L.: SARS-CoV-2 detection in air samples from inside heating, ventilation, and air conditioning (HVAC) systems- COVID surveillance in student dorms. Am. J. Infect. Control. **50**(3), 330–335 (2022). https://doi.org/10.1016/j.ajic.2021.10.009

50. Spinelle, L., Gerboles, M., Kok, G., Sauerwald, T.: 02 – sensitivity of VOC sensors for air quality monitoring within the EURAMET key-VOC project. Proceedings, 6–9. https://doi.org/10.5162/4EuNetAir2015/02 (2015)
51. Syriopoulos, T., Tsatsaronis, M., Gorila, M.: The global cruise industry: financial performance evaluation. Res. Transp. Bus. Manag. **45**, 100558 (2022). https://doi.org/10.1016/j.rtbm.2020.100558
52. Temirel, M., Yenilmez, B., Tasoglu, S.: Long-term cyclic use of a sample collector for toilet-based urine analysis. Sci. Rep. **11**(1), 2170 (2021). https://doi.org/10.1038/s41598-021-81842-z
53. Tseng, C.-C., Kung, C.-T., Chen, R.-F., Tsai, M.-H., Chao, H.-R., Wang, Y.-N., Fu, L.-M.: Recent advances in microfluidic paper-based assay devices for diagnosis of human diseases using saliva, tears and sweat samples. Sensors Actuators B Chem. **342**, 130078 (2021). https://doi.org/10.1016/j.snb.2021.130078
54. Weng, X., Neethirajan, S.: Aptamer-based fluorometric determination of norovirus using a paper-based microfluidic device. Microchim. Acta. **184**(11), 4545–4552 (2017). https://doi.org/10.1007/s00604-017-2467-x
55. Widjaja, S., Cohen, S., Brady, W.E., O'reilly, K., Susanto, Wibowo, A., Cahyono, Graham, R.R., Porter, K.R.: Evaluation of a rapid assay for detection of Chlamydia trachomatis infections in outpatient clinics in South Kalimantan, Indonesia. J. Clin. Microbiol. **37**(12), 4183–4185 (1999). https://doi.org/10.1128/JCM.37.12.4183-4185.1999
56. World Health Organization. (2007). International medical guide for ships : including the ship's medicine chest. World Health Organization, Geneva
57. World Health Organization: Meeting report – malaria rapid diagnosis – making it work. https://apps.who.int/iris/handle/10665/208030 (2023)

Technology Approaches for Cruise Ship Disease Propagation Monitoring

Bruno Almeida, Márcio Mateus, Miguel Pacheco, Pedro Maló, and Tiago Teixeira

1 Symptoms Detection

The first and most effective step on preventing a contagious outbreak is to encourage everyone to report their symptoms as soon as they are detected in order to control transmission chains. However, since many patients start transmitting the disease prior to any obvious symptoms, even if they report them, it may be too late to prevent further contamination.

Therefore, different approaches to early symptoms are described. With the increasing number and variety of sensors, combined with the vulgarization of IoT technologies, the range of detectable symptoms keeps increasing.

The first step was the definition of the relevant symptoms to monitor. For that, some of the most recent contagious outbreaks were considered, i.e., Norovirus, Influenza, and Covid-19. Another thing to note is the ability to monitor those symptoms with sensors, since some of them cannot be measured easily. With that, three symptom types were selected, namely fever, respiratory, and digestive symptoms.

1.1 Fever

First, the indicators taken into account for fever identification must be defined. The three most common measurable fever indicators were selected, namely body temperature, respiratory rate, and heart rate [1].

B. Almeida (✉) · M. Mateus · M. Pacheco · P. Maló · T. Teixeira
UNPARALLEL Innovation, Almada, Portugal
e-mail: bruno.almeida@unparallel.pt; marcio.mateus@unparallel.pt; miguel.pacheco@unparallel.pt; pedro.malo@unparallel.pt; tiago.teixeira@unparallel.pt

S. Th. Rassia (ed.), *The Blue Book*, https://doi.org/10.1007/978-3-031-48831-3_10

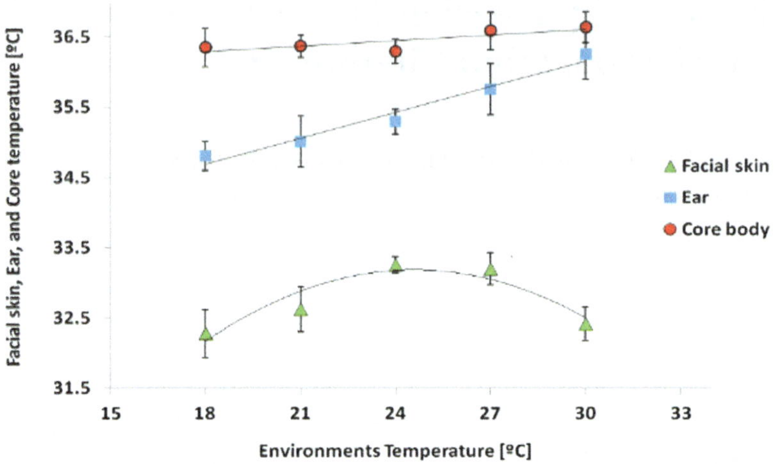

Fig. 1 Measured values of facial skin and ear temperature in comparison with the actual core body temperature [1]

Fever itself corresponds to an abnormal increase of the body core temperature [2]. It can be measured on different body areas with different levels of accuracy (Fig. 1). The problem is the most accurate measuring methods are invasive, which, in a public place like a cruise ship, is not viable, so an accurate yet contact-free solution must be adopted [2, 3]. One option stands out as the most studied, tested, and, therefore, proven viable in terms of both accuracy and practicality, this being infrared thermal imaging.

A body temperature rise may be a sign of fever but, alone, can be an inaccurate indicator, as many other factors may increase its value (e.g., ambient temperature or the menstrual cycle, in females) [2, 3]. In Fig. 2 is shown the effect of the ambient temperature on the temperature measured in two different skin areas as well as their discrepancy from the actual core body temperature. It shows that any external measuring technique will require a calibration process, depending on the body part where the measurement is taken.

Increased respiratory rate is also related with fever symptoms. The study conducted by Kapoor et al. links an increased breathing rate with 37.1% of the 1500 febrile participants [3]. The monitoring of the breaths per minute may, therefore, be helpful to early identify a fever case [2]. Like the body temperature, the solution for respiratory rate monitoring must be non-invasive. The same applies to the heart rate, even though fever increases its value, many factors like stress and fatigue also affect its normal value.

All the indicators described above face the same challenge, their values alone, i.e., the detection of an increase above a certain threshold may introduce false negative or positive results, since the normal value is not the same for everyone. This way, any approach to this may address this problem by finding a way to know each person's normal values [2].

1.1.1 Thermal Camera-Based Screening

This approach requires that each place of interest is provided with one or more thermal cameras that live monitor the skin temperature of the incoming people, triggering an alert for each measurement above the defined limit. Thermal cameras detect the infrared radiation associated with the heat to create an image with the estimated surface area of any object or body. The emissivity of a body is influenced by environmental factors, such as temperature, meaning that cameras need to be calibrated in order to map a certain amount of emitted radiation to the correct temperature. Because of that, many camera models come with a calibration reference that, when installed in the camera's field of view, allows for real-time calibration, since the reference temperature is measured *in loco* [4].

Most common models also include a visible optical sensor in order to blend the thermal info with a standard image and allow the recognition of the heat source. This, in combination with a recognition algorithm, may help filter false triggers. Multi-person, real-time continuous screening is also possible, due to the improved processors included, and, with that, many fever screening systems have been proposed [5, 6].

Two on-ship scenarios were considered: screening only upon embarkation and throughout the ship. The first one requires a simpler infrastructure but limits the screening rate to the entrance and exit of the ship and may lack accuracy [2]. The second option allows for a better understanding of the health state across the ship, since it is possible to install cameras in the most crowded areas. To identify a potential febrile person, facial recognition is required. Without it the outbreak control is limited to statistic indicators and no transmission chain control is possible. The main challenge of this approach is the authorization for facial recognition, which may raise privacy questions.

1.2 Personal Wearable Health Data Access

With the ever-growing popularity of wearable devices, some important health data are already being collected and made accessible by the major mobile health apps. One possible approach is to access these data and use it in combination with the thermal camera images to create a more accurate data set and, therefore, a more accurate fever indicator.

Nowadays, pretty much every model measures health data in order to provide indicators, warnings, and recommendations, namely heart rate, breathing rate, blood oxygen, mobility metrics such as steps that allow for an estimation of burned calories, among many others. Some models even include medical-grade sensors, so accuracy will not be an issue in most cases.

One less common feature is temperature measurement; however, with Apple's recent addition of skin temperature sensors to the Apple Watch 8, the market may follow, turning it into a standard. Since it is not a thermometer, it only measures shifts on the normal body temperature, which is measured overnight.

In terms of accessibility, the two major health apps, Google Fit and Apple HealthKit, provide APIs to access personal data if user permission is granted. Google Fit uses REST API for access to data, provided an access token is obtained. The data structure as well as all the available fields is all documented and publicly available [7]. The Apple HealthKit framework provides a set of queries that return the requested data from the HealthKit data store to authorized users [8].

The main challenge with this approach is the uncertainty of the data source. It is difficult to forecast the adherence to this approach either because of the number of smart wearable device users or their permission to access personal data. To analyze the viability of this approach, it is crucial to understand how a possible lack of data from some users affects its effectiveness.

1.3 Respiratory Symptoms

Respiratory symptoms are more difficult to monitor in a non-invasive way. Like mentioned in the previous section, wearable devices estimate some of these symptoms based on data from their built-in sensors, e.g., respiratory rate, but these are, in most cases, admittedly estimates and should not be used as medical indicators. Event detection like cough, sneezing, or nose blowing requires sound monitoring in order to detect its occurrence and analyze it or calculate its rate [9, 10]. The symptoms taken into consideration are:

- Cough
- Sneeze
- Nose blowing
- Respiratory rate

Apart from the respiratory rate, addressed in the previous section, all these symptoms may be monitored with any type of sound capturing device (provided it matches the quality requirements) and an algorithm that, with input audio signal processing and a large data set, can identify the events and filter any false trigger that may occur [11–13].

One challenge to this approach is the availability of data sets for all the symptoms considered. Some generic sound data sets available online include a large number of respiratory sound samples, for example, Audio Set, a data set created from YouTube videos that, among others, includes a variety of human sounds labelled accordingly [14]. Another example is the freesoundLabs data set that provides a variety of sound data sets, including human respiratory sounds [15]. While these may be helpful for event detection and false trigger filtering, they lack the specific disease element that may allow the detection of diseases through audio analysis. In that regard, and boosted by the Covid-19 pandemic, some specific cough/sneeze data sets have been generated, like CoughVid [16]. Pfizer even financed a challenge to boost the development of cough detection smartphone apps [17–19]. The same applies to nose blowing detection, with the added challenge that no specific data sets were found; however, the detection and rate measuring may be enough.

Just like for fever screening, one symptom alone may not constitute a good enough indicator. This approach requires that an extra layer of analysis from all the sources is implemented. Another challenge may be the permission to collect audio samples that, even if used exclusively for this purpose, may raise questions.

1.4 Digestive Symptoms

Digestive symptoms are the most difficult to monitor in a non-intrusive way. The proposed approach relies on toilet flush monitoring and possible digestive disease detection based on abnormal toilet usage. This can be achieved, for example, with the help of a flush sensor like shown in Fig. 2 or via flush sound monitoring.

Since a variation in toilet usage may be caused by multiple factors, namely the number of users, these data can only be used as an early indicator to aid in a more thorough symptom diagnosis. This way, sensors such as the Telehealth Flush Sensor [20] or a microphone could be installed in private toilets, for a more restricted monitoring, or in public toilets, to provide a general indicator of the toilet usage. This way, it is possible to detect deviations from the normal usage pattern that could trigger an alert and require another, more complete screening.

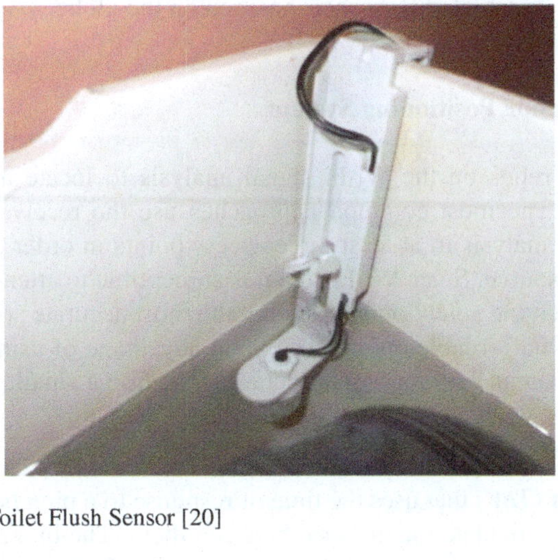

Fig. 2 Telehealth Toilet Flush Sensor [20]

2 Location of People

Since one of the main goals is to provide a way to quickly identify disease spread, locating each person inside the ship may be a powerful tool to track contacts. The approach to go for may depend on the accuracy required, i.e., how much data are needed for a robust contagion detection system and how important these data are for that detection.

In this regard, two different location approaches are proposed. The first one is a real-time location system that relies on continuous data acquisition from a personal device. The second is a contact tracing system to monitor exposure and possible contagion. Both approaches rely on a personal device and a communication protocol.

2.1 Real Time

GPS relies on a direct path between the device and one of the satellites, which means that using only GPS signal will not work indoors. What mobile devices use to solve this is a combination of GPS signal, the location associated to the Wi-Fi router, and the cell towers to which the phone is connected. When the ship is moving, since the Wi-Fi router location is static and registered upon configuration and cell phone signal is not guaranteed, location would be restricted to the outdoor areas of the ship.

To avoid this, alternative location systems are proposed. These approaches rely on signal analysis to locate a device relative to a fixed point or points on the ship, which is completely isolated from their position on the globe.

2.1.1 Wi-Fi Indoor Positioning System

This technology relies on the Wi-Fi signal analysis to locate a device in a Wi-Fi-covered area. The most common approaches use the received signal strength indicator (RSSI) analysis to at least three access points in order to triangulate and locate the signal source. Since Wi-Fi signal is susceptible to attenuation, reflection, or interference, relying solely on RSSI is not the most accurate solution. It may be, however, the most affordable option. The main advantage of using Wi-Fi signal is that it may work on an existing infrastructure or require a small, low-cost upgrade [21, 22] (Fig. 3).

More accurate yet complex techniques include:

- **Time-of-flight (ToF)** that uses the time of response to a ping between devices in the network to calculate the distance between them. The biggest challenge is to deal with the data synchronization between devices [21, 23].

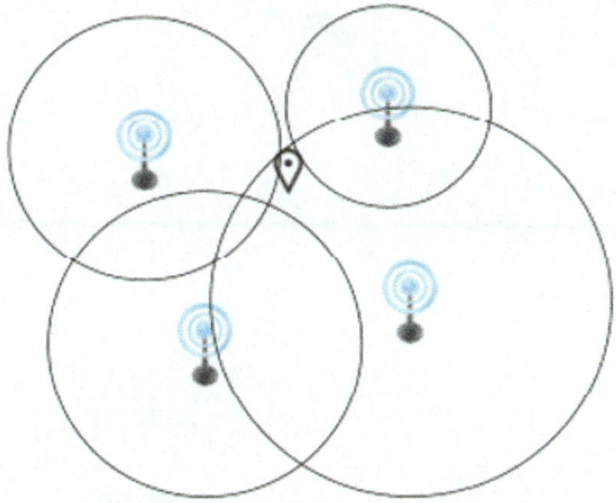

Fig. 3 Wi-Fi RSSI location system [22]

- **Angle of arrival (AoA)** that uses the phase shift of the incoming signal to calculate its angle of arrival and triangulate the source location. This is usually the most accurate technique but requires the largest number of access points [21, 23].
- **Fingerprinting** technique that works like de RSSI with the advantage that it stores all the data in a database, creating an indoor map based on signal strength. This technique provides higher accuracy but requires a long calibration procedure to populate the database. Another disadvantage is the fact that any changes in the environment, like moving objects, will affect the received signal strength and, therefore, induce location errors [21, 22].

2.1.2 BLE Indoor Positioning System

This location system uses Bluetooth instead of Wi-Fi signal. The most adopted location technique is RSSI; however, any of the above described may be implemented. Bluetooth Low Energy (BLE) and Wi-Fi both operate on the 2.5 GHz frequency, which gives them similar propagation properties, yet BLE is much more optimized for low power consumption. On the other hand, it requires a whole new infrastructure just for location purposes, while the Wi-Fi approach may rely on the existing infrastructure [22, 24].

This approach can be implemented using sensors that passively detect transmissions from any Bluetooth device, or with beacons, that continuously send pulse signals that can be detected by any Bluetooth device. The devices to be tracked may range from a personal device, like a smartphone, or a wearable beacon that could be, for example, a bracelet, a card, or a keychain. Some models integrate NFC, which may provide additional functionalities, such as contactless payment or access control [24].

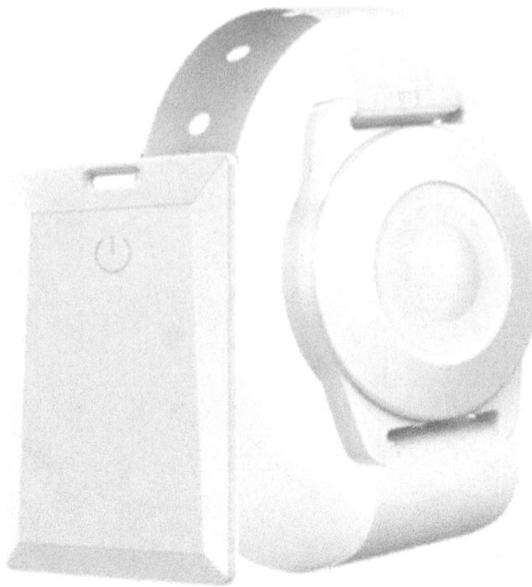

Fig. 4 Moko Smart Bluetooth tags in card and disposable wristband format [25]

2.2 *Contact Tracing*

This approach relies on Bluetooth signal to detect proximity between two personal devices. If that proximity exceeds the defined safe time, a contact is registered on a database, containing both device's IDs. Then, if one of these users shows any symptoms or is identified as ill, an alert is triggered and the whole contact chain can be identified. This can be accomplished using wearable devices such as the beacons shown in Fig. 4. However, empowered by the Covid-19 pandemic, a smartphone exposure notification system was created in a collaboration between Apple and Google to help health authorities control contagion chains by providing exposure notifications to any smartphone user that authorizes it [26, 27].

This system uses smartphone's Bluetooth to share a random ID (that changes every 15–20 min) with the surrounding devices. If contact is detected for more than the determined time, the devices share the random ID and the contact is stored in a database for 14 days. If, in that period, one of the users voluntarily identifies as infected, all the devices in the contact list are notified, after that all data are erased [26, 27].

By randomly updating the IDs and encrypting the exposure keys, no personal data, including ID, are exposed. Additionally, access to this service was only granted to official health authorities for app development; however, third-party and even open-source alternatives have been developed; therefore, a cruise ship implementation is possible.

References

1. Sun, G., Abe, N., Sugiyama, Y., Nguyen, Q.V., Nozaki, K., Nakayama, Y., Takei, O., Hakozaki, Y., Abe, S., Matsui, T.: Development of an infection screening system for entry inspection at airport quarantine stations using ear temperature, heart, and respiration rates. In: 2013 35th Annual International Conference of the IEEE Engineering in Medicine and Biology Society (EMBC), pp. 6716–6719 (2013)
2. Smarr, B.L., Aschbacher, K., Fisher, S.M., Chowdhary, A., Dilchert, S., Puldon, K., Rao, A., Hecht, F.M., Mason, A.E.: Feasibility of continuous fever monitoring using wearable devices. Sci. Rep. **10**, 21640 (2020)
3. Kapoor, M., Kumar, N., Panda, P.: The symptomatology of fever: a step towards qualitative definition of fever. J. Family Med. Primary Care **11**, 5990 (2022)
4. Nguyen, A.V., Cohen, N.J., Lipman, H., Brown, C.M., Molinari, N.A., Jackson, W.L., Kirking, H., Szymanowski, P., Wilson, T.W., Salhi, B.A., Roberts, R.R., Stryker, D.W., Fishbein, D.B.: Comparison of 3 infrared thermal detection systems and self-report for mass fever screening. Emerg. Infect. Dis. **16**, 1710 (2010)
5. Nishiura, H., Kamiya, K.: Fever screening during the influenza (H1N1-2009) pandemic at Narita International Airport, Japan. BMC Infect. Dis. **11**, 1–11 (2011)
6. Wei, P., Yang, C., Jiang, X.: Low-cost multi-person continuous skin temperature sensing system for fever detection: Poster abstract (2020)
7. Getting started with the REST API
8. HealthKit
9. Birring, S.S., Fleming, T., Matos, S., Raj, A.A., Evans, D.H., Pavord, I.D.: The Leicester cough monitor: preliminary validation of an automated cough detection system in chronic cough. Eur. Respir. J. **31**(5), 1013–1018 (2008)
10. Vizel, E., Yigla, M., Goryachev, Y., Dekel, E., Felis, V., Levi, H., Kroin, I., Godfrey, S., Gavriely, N.: Validation of an ambulatory cough detection and counting application using voluntary cough under different conditions. Cough **6**(1), 1–8 (2010)
11. Tena, A., Clarià, F., Solsona, F.: Automated detection of covid-19 cough. Biomed. Signal Process. Control **71**, 103175 (2022)
12. Larson, S., Comina, G., Gilman, R.H., Tracey, B.H., Bravard, M., López, J.W.: Validation of an automated cough detection algorithm for tracking recovery of pulmonary tuberculosis patients. PLoS One **7**, e46229 (2012)
13. Kruizinga, M.D., Zhuparris, A., Dessing, E., Krol, F.J., Sprij, A.J., Doll, R.J., Stuurman, F.E., Exadaktylos, V., Driessen, G.J., Cohen, A.F.: Development and technical validation of a smartphone-based pediatric cough detection algorithm. Pediatr. Pulmonol. **57**(3), 761–767 (2022)
14. Gemmeke, J.F., Ellis, D.P.W., Freedman, D., Jansen, A., Lawrence, W., Moore, R.C., Plakal, M., Ritter, M.: AudioSet: an ontology and human-labeled dataset for audio events. In: 2017 IEEE International Conference on Acoustics, Speech and Signal Processing (ICASSP), pp. 776–780 (2017)
15. freesoundlabs datasets
16. Orlandic, L., Teijeiro, T., Atienza, D.: The COUGHVID crowdsourcing dataset, a corpus for the study of large-scale cough analysis algorithms. Sci. Data **8**, 156 (2021)
17. Hale, C.: Pfizer to drop $74m for COVID cough-screening smartphone app developer (2022)
18. Melek, N.: Responding to challenge call of machine learning model development in diagnosing respiratory disease sounds (2021)
19. Patel, S., Rivas, A., Psaltos, D.: Dataset of sounds of symptoms associated with respiratory sickness (2020)
20. Toilet flush sensor
21. Wi-Fi location tracking and positioning
22. Mendoza-Silva, G.M., Torres-Sospedra, J., Huerta, J.: A meta-review of indoor positioning systems. Sensors **19**, 4507 (2019)

23. Ahmed, A., Arablouei, R., Hoog, F., Kusy, B., Jurdak, R., Bergmann, N.: Estimating angle-of-arrival and time-of-flight for multipath components using WiFi channel state information. Sensors **18**, 1753 (2018)
24. Bluetooth location tracking and positioning
25. Moko smart - products
26. Privacy-preserving contact tracing—Apple and Google
27. Exposure notifications: Helping fight Covid-19

Scenarios for Sensors and Actuators Deployment to Prevent and Mitigate Epidemics on Cruise Ships

Bérengère Lebental, Kaitlyn West, Olga Vainer, Michel Hoffman, Stathes Hadjiefthimiadis, Babis Andreou, Marc Bonazountas, Iro Palaiokosta, Astrinos Papadakis, Nasos Grigoropoulos, Amalia Ntemou, Anne-Sophie Fölster, Dimitris Drikakis, Bruno Almeida, Márcio Mateus, Miguel Pacheco, Pedro Maló, and Tiago Teixeira

1 Introduction

Since 2020, the heavy toll of the global COVID-19 pandemic on the cruise industry has been well documented. Highly publicized outbreaks on cruise ships in the early days of the pandemic were followed by reported cases of the virus being widely spread in passengers' home countries by those returning from cruise ships. By the end of March 2020, cruises were halted worldwide, and travel restrictions increased. This ultimately reduced the number of cruise passengers from 30 million in 2019

B. Lebental (✉)
Université Gustave Eiffel, COSYS, IMSE, Marne-La-Vallée, France
e-mail: berengere.lebental@univ-eiffel.fr

K. West
COLUMBIA blue, Hamburg, Germany

O. Vainer · M. Hoffman
Ecosense Group (Middle East) Ltd, Beit Yehoshua, Israel

S. Hadjiefthimiadis · B. Andreou · M. Bonazountas
Epsilon Malta Ltd, Swatar, Malta

I. Palaiokosta · A. Papadakis
Hydrus Engineering SA, Athens, Attica, Greece

N. Grigoropoulos · A. Ntemou
Netcompany-Intrasoft S.A., Research & Innovation Development Department, Luxembourg, Luxembourg

A.-S. Fölster
RWO GmbH, Gerold-Janssen-Strasse 2, Bremen, Germany

D. Drikakis
University of Nicosia, Nicosia, Cyprus

B. Almeida · M. Mateus · M. Pacheco · P. Maló · T. Teixeira
UNPARALLEL Innovation, Caparica, Portugal

S. Th. Rassia (ed.), *The Blue Book*, https://doi.org/10.1007/978-3-031-48831-3_11

to 6 million in 2020. It is estimated that cruise lines lost $77 billion between March and September 2020 alone [36, 51].

Meanwhile, the fight against COVID-19 on land has been bolstered by the use of the Internet of Things (IoT), and its myriad of sensors and actuators [7]. But within the cruise ship industry the use of these solutions has been and remains minimal. Accordingly, recommendations for managing COVID-19 on board cruise ships have been published and revised by various international agencies (such as European Medicines Agency) in recent years, but they do not include any IoT solutions [17], whereas such solutions have been widely heavily used by land-based agencies.

One of the goals of the Horizon Europe project Healthy Ship for You (HS4U[1]) is to evaluate how the current advances in the field of IoT, especially in the area of healthcare sensors and actuators (S&A), could be applied to the prevention and mitigation of outbreaks on board large passenger vessels, particularly cruise ships and ROPAX (roll-on/roll-off passenger) vessels.

To evaluate the feasibility of achieving this high-level objective, the consortium formulated a series of high-level technical, operational, and business requirements. It then analyzed the state of the art to identify a number of promising S&A-based technology solutions that meet these high-level requirements. It resulted in scenarios for S&A deployment aboard cruise ships. The present chapter reports on the highlights of this process.

2 High Level S&A Requirements

2.1 Outbreak-Related Requirements

Relevant outbreaks on cruise ships have been identified by industry stakeholders in documents such as SHIPSAN [16] and the US Vessel Sanitation Program [54]. These diseases include gastroenteritis, influenza-like illness, Legionella and vaccine-preventable diseases (e.g., measles, rubella and varicella).

To address such diseases on cruise ships, the HS4U project listed two main objectives for S&A:

1. The first objective aims to detect an outbreak, i.e., to detect as early as possible any person on board with an infectious disease. This breaks down into two steps: detection of a sick person and identification of the actual disease. The latter (i.e., not just the detection of symptoms) is particularly critical for initiating the appropriate outbreak mitigation measures, as these are disease-specific. To detect patient zero, exhaustive monitoring of passengers and crew members is required. To minimize the number of contact cases after initial contamination, detection as close as possible to the onset of symptoms, or even before, is needed.

[1] https://hs4u.eu/, last accessed May 24th, 2023.

Considering that health-related data is to be collected, extreme care should be taken with regard to data protection. Non-anonymized health-related data should remain confidential throughout its life cycle. People who may have access to such data, beyond the ship's medical personnel, should be carefully defined based on the regulatory framework.

2. The second objective aimed at outbreak prevention and mitigation. One aspect directly targets pathogens: prevention of pathogen colonization (on surfaces and/or in piping systems) and transmission (through air, water or surfaces, including piping systems). This requires solutions for permanent and/or continuous and/or automatic disinfection of surface, air and piping. The second aspect aimed to minimize exposure of people on board: detecting exposure by direct contact or by staying in a contaminated area; enabling confinement procedures; enabling smooth information transfer (between crew, passengers, and authorities).

2.2 Business Requirements

The business requirements of the cruise ship stakeholders (ship operators, ship owners, ship hotel managers) were captured through interviews and workshops. The companies *Celestyal Cruises* and *COLUMBIA blue* are members of the HS4U consortium and contributed to the following high-level requirements.

1. The cruise ship industry is a luxury business. Priority must be given to passenger acceptability of the proposed solutions. This includes respecting and anticipating ethical and privacy concerns. To ensure acceptability, dual use of the solutions must be considered. For example, a smart wearable device used for outbreak monitoring can be positively marketed as a self-health tool provided by the cruise line; or it can be used to replace the key cards for room and restaurant access.
2. The cruise industry is also a mid-volume market: passengers number in the thousands, so the cost per passengers must be highly controlled and the solutions selected must be commercially viable in the thousands.
3. To limit costs, the solutions should require minimal additional crew effort to operate and maintain the systems.
4. As far as possible, the solutions should be transferable from ship to ship (i.e., when a ship changes ownership or is retrofitted, most technologies should be deployable to another ship). This implies the interoperability of the solutions, i.e., the ability to adapt them smoothly to different ship IT infrastructures.

Quantifying the affordability of the costs remains an open discussion. As seen in the case of COVID-19, in the event of a large outbreak on board, all passengers may have to be sent home and the ship's operations may have to be halted for several days or weeks. This can cost the ship's owners millions of dollars, not to mention the damage to their reputation. But, with the exception of the COVID-19 crisis (an extreme event), this is a rare scenario.

On the other hand, from a legal perspective, ship stakeholders are only required to implement on board the health-related measures defined in the applicable regulatory documents. Determining the affordability of implementing additional outbreak prevention and mitigation measures beyond those required by regulations is a risk and liability management issue. It seems reasonable to assert that the cost per passenger should not exceed a small percentage of the revenue per passenger. To put concrete numbers on this, the website cruisemarketwatch.com[2] reports that, in average, in 2021, passengers spent $214 per day on cruises with an average length of 7 days. The same source also reports the breakdown of this spending into operator expenses, from 14.5% of various operating costs down to 3% interest expense, including 11% of ship fuel costs or payroll, resulting in 12.7% profit before tax. Setting aside 0.5–1% of the profits – $1–2 per day per passenger – to insure against onboard outbreaks (which can cause millions of dollars in losses) can be argued as a reasonable amount compared to other expenses. Projected over a year, this amounts to an annual budget of approximately $500,000 per year for every 1000 passengers.

3 Ship Integration Constraints

3.1 Overview

Any S&A solution proposed for deployment on board must comply with existing rules and practices for ship design. All vessels must be constructed in accordance with prevailing rules and regulations to ensure their safe and secure operation throughout their lifetime. Drivers of legislation, both national and international, include statutory requirements (i.e., legal requirements), classification societies, flag states, international standards, and port authorities.

Cruise ships are specialized passenger vessels, with no cargo deck, designed for the commercial transportation of passengers in overnight accommodations. They can carry a large number of passengers and crew members, both for ship operations and hospitality, making safety on board of paramount importance.

A ship is made up of multiple systems into which each S&A solution must seamlessly integrate. A ship is a *system of systems*, consisting of various smaller systems that work together to achieve the ship's function. They are generally categorized as hull and machinery; deck, engine, and hotel machinery and equipment; engine propulsion and power generation systems; auxiliary systems for cooling, fuel, and lubrication; systems for handling waste and sewage; systems for providing hot and cold potable and recreational water; firefighting and life-saving equipment; and automation and communication systems.

[2] https://cruisemarketwatch.com/financial-breakdown-of-typical-cruiser/, last accessed on June first, 2023.

All the subsystems are integrated within the spatial arrangement of the vessel. Successful system integration requires a holistic understanding of the ship as a system so that no subsystem negatively interferes with another. Each system may consist of piping components and valves, tanks, electrical and mechanical equipment, etc., that must be coordinated. The engineering design is guided by the relevant regulations, the functional requirements of the system, and the needs of the customer. Based on the specific functionality that the system has to perform, the marine engineer plans for its interconnections and arrangements. All of the necessary information is included in the ship's drawings. Some of the most basic drawings are the general arrangement plan (i.e., GA plan), the capacity plan, the deck arrangement plans, the engine room machinery and electrical arrangement plans, the electric load analysis, the wiring diagram of power, the wiring diagram of communications, the wiring diagram of alarm and monitoring, the wiring diagram of lighting, the fire control plan, the piping diagrams.

With regard to epidemic management on board with S&A, the following systems are relevant for further discussion: piping and HVAC systems, as they are avenues of disease transmission, can contain and grow pathogens, and can be included in outbreak mitigation protocols; power generation systems, since the S&A must be powered; alarm and monitoring systems, to control and monitor the S&A; communication systems, to ensure that the S&A do not interfere with critical communications, and for use by the S&A for communication.

3.2 Piping Systems

Focusing on the piping systems on board, the systems of most interest to epidemic management are: fresh and potable water systems (including desalination plant, if applicable); ventilation systems; sanitary system. Note that the fire and safety system is not directly relevant to epidemy management, so is not listed here, but a similar approach to fire zones could be used for virus containment.

A piping and instrumentation diagram, or P&ID, shows the piping and related components of a physical process flow. Various operational scenarios are integrated into the piping system. For each, the appropriate valve arrangements, types, and instruments are selected. Valves are used in a piping system to regulate or stop the liquid flow. Various types exist onboard a vessel, like butterfly valves, globe valves, ball valves, gate valves, non-return valves, etc. The valves may be operated either manually by the crew or remotely through a control station. The open/close status of some manually operated valves can be displayed at a control station. The main control stations are located in the Engine Control Room (ECR), the Engine Control Console (ECC), and in the Wheelhouse, the Bridge Control Console (BCC). Remote control is generally enabled by electrical, pneumatic, or hydraulic means. The various on-board sensors and all actuation means are shown in the P&IDs.

The piping system is usually monitored for pressure, temperature, and flow. The type, number, and localization of monitoring instruments are defined based on the ship's structural design and on the functionality of the specific system. The signals from the instruments, valves, and equipment are usually gathered at electrical panels that are connected with the Alarm and Monitoring system of the ship.

In this context, S&As for outbreak management are a new functionality that needs to be integrated into the ship's design. Any piping system can either be retrofitted with S&A or retrofitted by changing part of its design, including new valves or restricted connections between piping systems for existing vessels. On new vessels, a piping system can be redesigned using the best practices identified from experience or the requirements of codes and regulations.

3.3 Gray and Black Water Piping System

There are two types of wastewater piping systems on board. The first is the *black water system*, which handles all of the sewage. As defined in MARPOL Annex IV,[3] "sewage" refers to water and other wastes from toilets and urinals; drainage from medical premises (dispensary, sick bay, etc.) through wash basins and scuppers located in such premises; drainage from spaces containing living animals; or other wastewater when mixed with the drains defined above. The second is the *gray water system*, which manages drainage from dishwashers, galley sinks, showers, laundry, baths, and washbasins, and does not include drainage from toilets, urinals, hospitals. With regard to early disease detection, the black water system is the most critical: high virus loads can be detected there in the early stages of an outbreak.

Ships use a variety of sewage disposal systems depending on their size and technology. Either gravity or vacuum toilets can be found. Sewage is discharged either into common sewer lines leading to the ship's side and overboard through storm valves, or into collection and storage tanks, with or without aeration facilities, in which the sewage is retained for disposal on land or at sea, or into sewage treatment systems incorporating a combination of collection and treatment tanks, with processes designed to eliminate organic and sanitary pollutants in the sewage into effluent suitable for discharge into the sea without harmful effects. Ship requirements for sewage handling are detailed in MARPOL Annex IV.

On large passenger ships, blackwater is typically collected in the holding or treatment tanks using vacuum collecting systems. Depending on the size and number of passengers, 4–10 vacuum collecting units are installed per ship. Due to increasing environmental awareness, large passenger ships are being equipped with Advanced Wastewater Treatment (AWT) Systems using biological sewage treatment and membrane or flotation technology. Even though MARPOL Annex IV does not

---

[3] International Convention for the Prevention of Pollution from Ships – Annex IV Prevention of Pollution by Sewage from Ships entered into force 27 September 2003. https://www.imo.org/en/about/Conventions/Pages/International-Convention-for-the-Prevention-of-Pollution-from-Ships-(MARPOL).aspx, last accessed June 11th, 2023.

regulate graywater, AWTs are often designed for the vessel's entire wastewater system, which consists of both black and graywater discharges. The high specific wastewater volumes, up to 350 L per person and per day, makes the installation of sewage treatment plants a preferred option on board large passenger ships: as the operators are allowed to discharge the treated effluent directly to the sea, the construction of large-capacity sewage holding tanks is avoided, which is valuable as space is a limited resource on board ships.

3.4 *HVAC System*

HVAC refers to the heating, ventilation, air conditioning, and cooling systems onboard a vessel. The ventilation system is a complex system with high power consumption. Temperature, pressure, and humidity must be monitored to ensure proper operation. Ship ventilation systems have different requirements depending on the destination of the compartments and the type of ship. For example, there are compartments where natural ventilation is sufficient. Other compartments require forced ventilation by means of exhaust and/or intake fans, while still other compartments require ventilation by means of an air handling unit. In the case of living quarters, standard ventilation and air conditioning systems are used to provide cooling or heating of the compartments and the required supply of fresh air. Air purifiers may also be used [15].

In a standard ventilation and air conditioning system (Fig. 1), fresh air supplied through openings in the vessel structure is mixed inside the Air Handling Unit (AHU) with air recirculated from the rooms it serves. At the inlet to the AHU the recirculated air can be mixed with the fresh air from outside to equalize the temperature difference between the hot and cold air. Based on the volume of the space and the mixing ratio at the AHU, the air is forced into the distribution piping that served the area. An AHU generally consists of a fan that draws air through a mixing chamber for fresh and re-circulated air, filters, heating coil (steam), cooling coil for evaporative refrigerant, and steam humidifier nozzles.

The air handling unit can supply air that has been heated or cooled. Only one function can be active at a time. When the outdoor temperature exceeds the required set point, the control valve to the heaters is closed and the cooling control system operates. When the outdoor temperature is below the set point, steam regulation is initiated by opening the control valve to the heaters and closing the valves in the refrigerant lines. A temperature sensor in the supply line controls the cooling and heating. Should the AHU fan stop or trip, the cooling and heating valves will automatically close.

In large ships, there are multiple AHUs on different decks to serve different areas, defining HVAC zones. An HVAC zone is a space or group of spaces that are independently controlled for temperature, humidity, and air distribution. Zones are directly related to fire zones and are grouped by cabin and public areas. A zone usually consists of common ductwork served by an air handler.

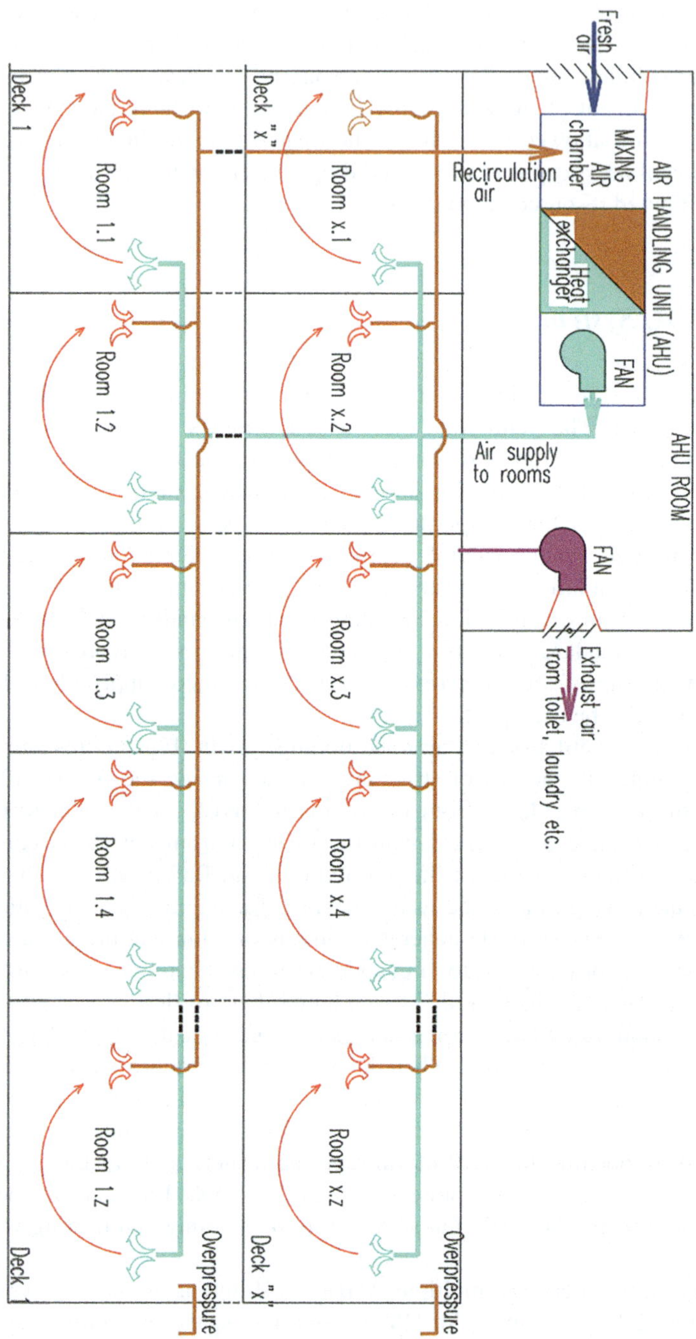

Fig. 1 Common ventilation and air conditioning system. (Reproduced from Mihai and Rusu [35])

Compared to the generic configuration described above, cruise ships have specifics: there are always supply and return ducts with two fans in passenger areas, plus an extraction fan from toilets. Cruise ships use chilled water systems in the state of direct refrigerants to AHU. The chilled water is produced centrally in the engine room and can thus provide cooling in other areas too (e.g., galley equipment).

4 State of the Art for S&A for Epidemic Management on Cruise Ships

4.1 The Internet of Ships

Based on the requirements listed in the previous section, the state of the art in S&A was investigated to identify suitable solutions. Overall, it appears that the field of Internet of Things (IoT) on ships is gaining traction. Initial research in this area was conducted by Paik et al. who detailed the use of wireless sensor networks as decision support tools on ships [38, 39]. Later, Aslam et al. proposed the "Internet-of-ships" paradigm, which includes ship applications of S&A such as safety enhancements, route planning and optimization, collaborative decision-making, automatic fault detection and preventive maintenance, cargo tracking, environmental monitoring, energy-efficient operations, and automatic berthing [3]. However, as of May 2023, the concept of the "Internet-of-ships" does not include healthcare applications: bibliographic searches combining "ship," "healthcare," and "sensor" (and their synonyms) remain entirely fruitless.

Still, more specific keywords yield interesting references. In particular, the COVID-19 pandemic inspired several studies: Mihai and Rusu reviewed in 2021 how optimizing the operation of the ship ventilation systems could prevent the spread of diseases [35]. In addition, Liu et al. [30] discussed people tracking for the mobile ship environment (to subsequently allow contact tracking) [30]. Furthermore, on a directly related topic, Liston et al. [29] addressed the acceptability and usability of location-support technologies for cruise ships, but applied this to the issue of ship evacuation [29].

Another driver for the use of IoT on ships is the "silver economy," i.e., the idea of using IoT to make cabins more suitable and accessible for elderly passengers using the IoT. This has been elucidated, for example, as a *smart cabin* outfitted to detect fall events [55]. Several alternative approaches have also been proposed, whereby the e-cabin provides insights into smart maintenance scheduling, energy usage optimization, and detection of security issue [5, 37, 44]. In these projects, the focus was primarily placed on comfort as opposed to healthcare.

4.2 Healthcare S&A: Applicability to Ships?

Considering the scarcity of the literature, HS4U endeavored to create an inventory of S&A technologies for outbreak prevention and management that could be adapted to the cruise ship context. With respect to sensor solutions for outbreak detection, the results of this inventory are reported in several chapters of this book: Almeida et al. discuss existing solutions for detecting symptomatic passengers [2] while Lebental describes solutions for identifying infectious diseases [27]. In terms of actuators to prevent and mitigate outbreaks, Almeida et al. also discuss the human-related tools: how to track contact cases and passenger localization on the ship, and how to share information between the crew and the public.

As for solutions to combat pathogens, transmission of pathogens in closed indoor spaces occurs by direct or indirect contact such as physical contact (e.g., handshake), fomite (e.g., clothing and furniture), droplets (5–10 μ), and aerosols (less than 5 μ) [13, 14, 58]. Many disease outbreaks can also be attributed to the presence of microbially contaminated surfaces. Microbes, for example, can thrive and multiply in a moist environment when attached to any artificial surface, often forming a biofilm that provides a protective environment for microbial survival under challenging conditions [32]. Viruses, on the other hand, do not replicate outside a living microorganism but can be easily transmitted via inanimate surfaces simply by settling on a surface. Thus, contaminated surfaces in close proximity to infected individuals can serve as a medium for the spread of infections from animate sources to others via contaminated inanimate surfaces [25]. An important example is *norovirus,* which is responsible for more than 90% of nonbacterial and about 50% of all epidemic gastroenteritis cases, causing an estimated 267 million infections worldwide each year. One of the main routes of transmission of norovirus is through contaminated surfaces, either by touching an infected person or by droplets of aerosolized vomit from an infected person.

Existing disinfection or contamination removal techniques typically rely on UV light and filters to control airborne pathogens or chemical disinfection to eliminate surface pathogens. In fact, surface treatment has shown to reduce the microbial and viral burden on surfaces, thereby reducing disease transmission and development [57]. However, conventional sanitization methods that rely on chemical agents have been shown to be inadequate in preventing recontamination of surfaces and can actually promote the selection of drug-resistant strains. As a result, more than 50% of surfaces remain persistently contaminated [6, 45].

With regard to air purification, Dbouk et al. discuss the role of air purifiers in reducing indoor virus transmission [15]. Filtration is a widely accepted method of removing airborne pathogens and is particularly effective when used in conjunction with UV light. The addition of filters to central ventilation systems is likely to reduce the airborne load of infectious particles. However, the costs associated with energy consumption and the need to periodically replace filter media and UV lamps make cost-effectiveness a critical factor to consider [23]. Furthermore, if not properly designed or operated, the HVAC system can actually accelerate the transmission of the airborne disease throughout whole buildings [19]. In addition, HVAC systems themselves can be a significant contributor to indoor biological pollution and cause disease outbreaks, as certain microorganisms thrive in the dust of air conditioning units that provide the ideal temperature and humidity conditions [31]. Overall, poor installation and low maintenance of HVAC systems is a significant factor in disease propagation and transmission due to high risk of infestation with dangerous pathogens such as *Aspergillus* spp., which can cause lung disease and allergic asthma, or *Legionella* spp., which can cause a severe pneumonia called Legionnaires' disease [4, 46].

On the whole, it seems that extensive sanitation has become the primary approach in the prevention and disruption of disease outbreaks, despite the risk of accelerating the development of multidrug resistant (MDR) microorganisms. In the search of new effective and sustainable approaches to infection control, HS4U brings to light two approaches: colloidal silver coatings and probiotics. The state of the art, potential and challenges of applying colloidal silver as an anti-pathogen material in the naval sector are discussed in the chapter by Somarriva et al. of the present book [49, this volume].

With regard to probiotics, their value in "attacking" surface pathogens is being actively explored [1]. The accumulation of studies on the effect of probiotics on pathogens on inanimate surfaces has shown that the introduction of certain probiotic strains can antagonize pathogen growth and reduce pathogen biofilm formation through mechanisms of competitive exclusion (competition for resources) [18]. Bacterial strains belonging to Bacillus genera are among the most commonly used probiotics, probably due to their ability to form spores, propagate in various environments, and are considered GRAS safe. Other factors contributing to the high potency of Bacillus spp. are the production and secretion of various antibiotic compounds and significant amounts of proteolytic enzymes with outstanding stability [8, 33, 43]. It has been shown that *Bacillus subtilis* strains expressing extracellular enzymes can effectively inhibit SARS-CoV-2 virus infection by degrading spike proteins [52].

Different formulations of surface cleaners based on probiotics containing Bacillus spp. have shown remarkable results when applied topically on surfaces compared to the common cleaners such as soap or 70% ethanol solution [11, 12, 50], in some cases showing a stable reduction of surface pathogens up to 90% more than conventional disinfectants, without selecting resistant species [6]. Despite its potential benefits, the use of probiotics for topical application also has some drawbacks similar to those of conventional disinfectants. These include being a time-consuming process, requiring frequent reapplication, and not being able to effectively reach all the small areas where microorganisms may hide and continue to proliferate. *Ecosense LTD*, a member of HS4U consortium, is addressing these issues as described in a later section.

5 HS4U Scenarios for S&A

5.1 Overview

Based on the state of the art and the high-level requirements described above and in the related chapters of this book, the following paragraphs detail the scenarios of sensors and actuators that HS4U deems likely to be realized in the near future for outbreak prevention and mitigation aboard cruise ships. The scenarios are illustrated in Figs. 2 and 3.

5.2 Sensors for Early Detection of Outbreaks

The ideal sensor system for early detection of outbreaks would allow for each person on board to be continuously monitored anywhere on the ship for symptoms

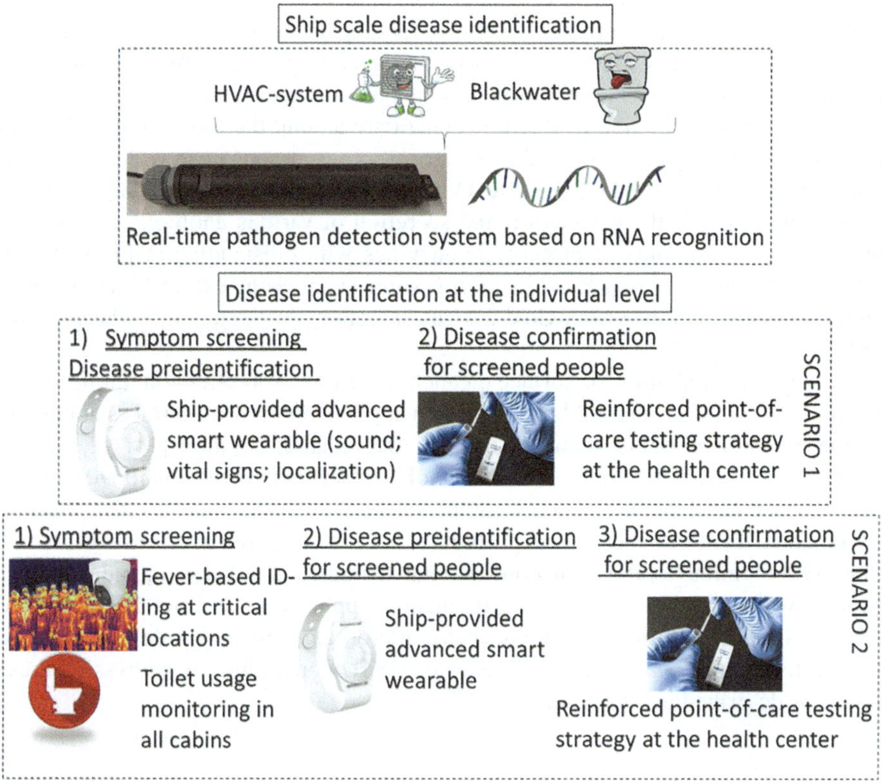

Fig. 2 Sensors for early detection of outbreaks

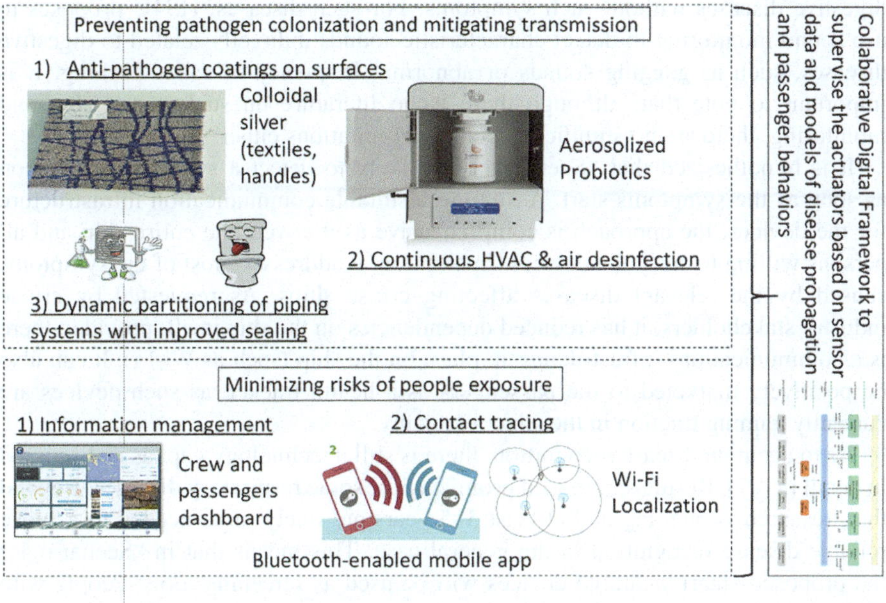

Fig. 3 Actuators for prevention and mitigation of outbreaks. (Bluetooth-enabled mobile app – image adapted from https://edps.europa.eu/data-protection/our-work/publications/techdispatch/techdispatch-12020-contact-tracing-mobile_en, last accessed June 29th, 2023. Wi-Fi localization – image reproduced from [34])

of relevant diseases. The disease should be detected before the person becomes contagious. The method should be selective and able to describe the specific diseases, not just that a person is sick.

The state of the art presented by Lebental [27] suggests that the best approach to achieve an outcome where diseases are identified is to equip each person on board with a cheap smart wearable device with the capability to monitor a predefined set of biomarkers and volatile organic compounds (odor) in sweat. There is no doubt that this will be technically feasible in the future, as the elementary building blocks of such a solution are already available, as evidenced in the literature. However, challenging research and development will be required to bring these components together at a commercial level, and even more effort will be required to reduce the cost of the final solution, while the market for such a device remains highly uncertain. Overall, it is estimated that it will be 5–10 years before such devices are affordable for the cruise industry.

In the meantime, alternative solutions should be proposed. “Scenario 1” is to have all passengers carry a smart wearable device monitoring vital signs and sound. According to the literature, such a device would be able to identify most respiratory diseases and some of the digestive diseases. This includes diseases that have fever or abnormal heart rate as symptoms. The gap in this approach is the detection of digestive diseases without such symptoms. For such diseases, HS4U proposes to use sound monitoring to detect characteristic sounds indirectly related to digestive diseases, such as gagging sounds or abnormal frequency of toilet flushing. It is important to note that, although there is no literature on such a use of sound monitoring, there are no significant technical limitations either.

It is hypothesized that "Scenario 1" is likely to detect a symptomatic person as soon as the symptoms start. Assuming a suitable communication infrastructure for the devices, the approach is comprehensive as it covers the entire ship and all persons willing to carry the devices. In addition, it addresses most of the symptoms caused by the relevant diseases affecting cruise ships. As requested by cruise industry stakeholders, it has reduced dependencies on the ship itself, provided there is a communications infrastructure in place on the ship (such as Wi-Fi). It can also be positively marketed to the passengers as a health tracker, as such devices are currently gaining traction in the silver economy.

As to accurate disease recognition, there is still a technology gap, as is discussed in detail in [27]. Results are indeed promising for some respiratory diseases. Beyond that, research is ongoing and at least 3–5 years are likely to be needed to achieve reliable disease recognition in the general case. This means that in "Scenario 1," the proposed smart wearable devices will be used as screening tools. People with abnormal symptoms will be referred to the ship's doctor for further evaluation free of charge. At the ship's health center, an enhanced point-of-care diagnostics offering, with a sufficient number of rapid tests for the most acute infectious diseases, will allow medical teams to quickly identify the disease and the patient's needs.

In terms of budget for "Scenario 1," a smart wearable of sufficient quality can be estimated to cost at least $30. Therefore, the total budget per 1000 passengers is estimated at $30,000. This budget is reasonable on an annual basis, not on a weekly basis, so the devices should be loaned to the passengers, not given away. Assuming 5% of the devices are lost each week, many devices should be completely replaced every 20 weeks, for a total cost of about $80,000 per year per thousand passengers. In comparison, the cost of rapid diagnostic tests is negligible (<$1 per unit for bulk orders) and is not included in the cost calculations. It should also be noted that the cost of "Scenario 1" can be reduced by optimizing the choice of devices. There are lower cost devices that may be sufficient for screening purposes; on the other hand, more durable devices (possibly more expensive as well) would allow for a better weekly return rate. Finally, passengers could be encouraged to take better care of the devices and to return them more reliably.

A total cost of $80,000 per year per 1000 passengers may be more than cruise ship stakeholders are willing to pay. In addition, some passengers may refuse to carry the devices or carry them only intermittently. This is expected to limit the overall performance of "Scenario 1."

An alternative solution, "Scenario 2," is proposed based on findings reported in [2]: thermal cameras would be used as fever-screening tools. They would be deployed in the most critical passageways of the ship. The suitable locations for such tools are those where passenger identification is possible, particularly in gangways and restaurants. For illnesses for which fever is not a symptom – mostly digestive disorders – each cabin would be equipped with a sensor monitoring the frequency of toilet flushing. This would have the added benefit of detecting malfunctioning toilets, resulting in improved passenger comfort and water savings. Passengers detected with a fever or abnormal toilet use would be offered the option of either going to the health center or wearing an appropriate smart wearable device (such as defined in "Scenario 1") for a period of time.

In terms of cost, this approach would be significantly less expensive: assuming $1000 per camera for 10 cameras with a lifetime of 3 years each; and 1 toilet sensor (<$5) per cabin for 500 cabins with also lifetime of 3 years, the total estimated cost would be less than $20,000 per year. However, this approach requires the wiring of hundreds of sensors, which would add significant installation costs (which are not easy to estimate in advance) to the equipment costs and significantly reduce the transferability of the solution from ship to ship. The cost of the fixed sensors must be buffered by a small number (approximately 50 units – 5% of a thousand passenger) of smart wearable devices to monitor symptomatic passengers identified by thermal cameras or by toilet sensors. These devices would be more durable than in "Scenario 1" because they would not be used systematically, but only on a case-by-case basis. Assuming a lifetime of 1 year in this scenario, the cost would be approximately 15,000$/year.

In total, "Scenario 2" (monitoring via thermal camera and toilet flushing) is about 50% less expensive than "Scenario 1" (wearable devices for all passengers), but significantly less effective because it detects only two symptoms (fever and abnormal toilet use) and detects these symptoms in specific locations, not throughout the ship. Passengers who are very active even being sick may mostly use public restrooms and not be detected as sick, while passengers with very low activity due to fever may not go to public areas and not be detected as symptomatic. Also, "Scenario 2" is more difficult to market positively to passengers compared to the smart wearable devices that can be used for self-healthcare.

In any case, both scenarios will be limited by the fraction of people who will refuse the use of the technologies described above. Partial coverage of the persons on board means that in the event of an outbreak, patient zero may be missed, which may exacerbate the outbreak. HS4U is conducting a survey of cruise ship passengers to assess the acceptability of scenarios 1 and 2 and to thus determine what fraction of the ship's population could actually be monitored.

To ensure early detection of the outbreak, even in the absence of full coverage of the ship's population, HS4U also proposes to deploy pathogen detection systems throughout the ship's HVAC and sewage systems. A pathogen detection system (PDS) is a real-time, continuous sensor that directly detects the pathogen or its RNA at the point of use. Such a solution would alert the crew that a disease may be circulating on board, ideally well in advance of the actual outbreak, allowing the onboard teams to initiate outbreak prevention and response plans. Depending on the density of sensors deployed, they could also provide information about the most infected areas, which would then be subject to enhanced disinfection protocols.

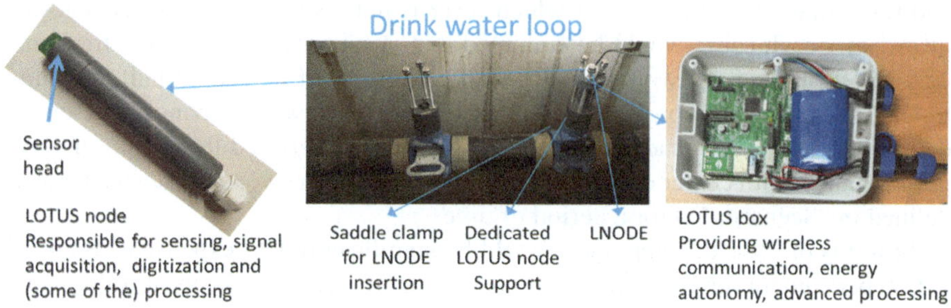

Fig. 4 LOTUS multiparameter water quality sensor deployed in a drink water look with wireless capabilities

Theoretically, PDS could be deployed in all the cabins, advantageously replacing the two scenarios described above. In practice, PDS are not available in commercial products today, regardless of cost [26, 41]. As discussed in [27], air, urine, and stool sampling systems are available, but automated, real time sample analysis for pathogen detection is not yet commercially available. What is available are solutions for rapid testing of samples by an operator, according to the modalities of point-of-care testing. Commercial availability of automated sampling and pathogen detection is expected to be 3–5 years in the future, at a cost likely to exceed \$1000 per sensor. Therefore, HS4U includes PDS in its scenarios, but in small numbers only: typically, one sensor would be deployed in each air handling unit of the HVAC system. Wastewater-based epidemic monitoring would be realized either on the inlet sampling point of the sewage treatment plant or by integrating further sampling points on the vacuum collecting units or in the black water pipes.

To accelerate the commercial availability of PDS, HS4U also includes the development of PDS for the HVAC system. The technology will be an iteration of the real-time, continuous sensor platform for water quality developed in the LOTUS project[4] by University Gustave Eiffel [10, 42] (Fig. 4). This sensor platform uses a functionalized carbon nanotube chemistor array for pH and chlorine monitoring (Fig. 5). Adding the ability to detect the RNA of pathogens to such a platform is a natural evolution of the technology. It requires functionalization of the carbon nanotubes with the appropriate RNA or DNA strands (Fig. 6), as is widely practiced in the literature [28]. These technologies are not necessarily reversible, which will be the main challenge to be overcome by Université Gustave Eiffel as part of the HS4U consortium.

[4] https://www.lotus-india.eu/, last accessed June 11th, 2023.

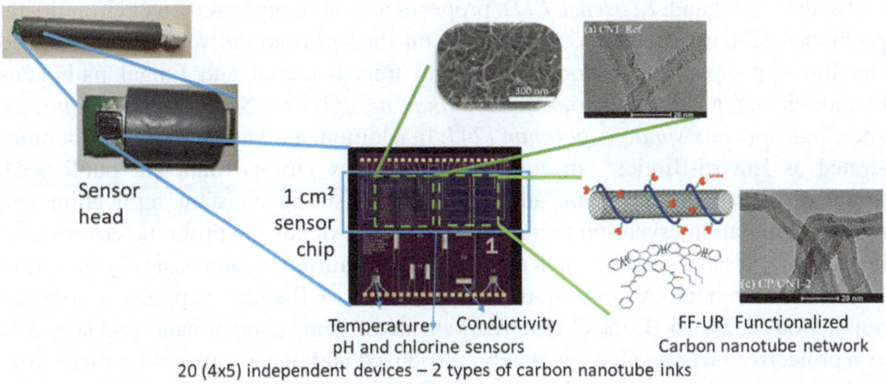

Fig. 5 Principle of operation of the LOTUS sensor: functionalized carbon nanotube chemistry array for pH and chlorine monitoring

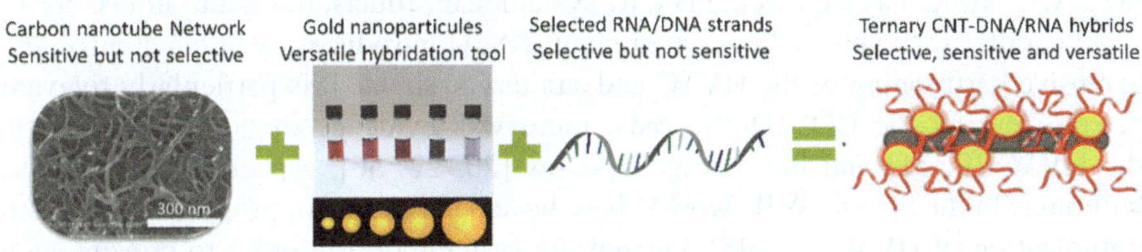

Fig. 6 Adaptation of the LOTUS sensor for DNA/RNA detection: The carbon nanotubes are functionalized with gold nanoparticles, themselves functionalized by DNA or RNA, forming a ternary hybrid nanomaterial. (Gold nanoparticles: copyright Aleksandar Kondinski – kondinski.webs.com. Ternary hybrid: Image reproduced from [53])

5.3 *Actuators for Prevention and Mitigation of Outbreaks*

For outbreak prevention and mitigation, HS4U's first approach is based on advanced anti-pathogen treatments designed to complement the regular hygiene standards applied by the crew throughout the ship. These solutions are generally designed to inactivate pathogens and are therefore relevant both to prevent the initial contamination by limiting the ability of pathogens to colonize, and to prevent further transmission from index cases by minimizing the density of pathogens. HS4U promotes two subtypes of anti-pathogen treatment described below.

The anti-pathogen treatment developed by *CNT Labs* uses thin coatings of colloidal silver to inactivate pathogens. More details are provided in A. Sommariva et al. [49]. Although various surfaces have shown a high density of pathogens during the COVID-19 crisis, such as sink countertops, handles, and floors [40], the HS4U approach targets textile surfaces such as carpets, drapes, and upholstery. Indeed, these surfaces are much more difficult and expensive to clean regularly than bathroom fixtures and door handles.

On the other hand, *Ecosense LTD.* proposes a technology based on environmental probiotics (Enviro-Biotics®). It is based on the airborne delivery of aerosolized Bacillus spp. spores to indoor surfaces to treat bacterial and fungal pathogens including, but not limited to, *Pseudomonas aeruginosa*, *Staphylococcus aureus*, and *Cladosporium sphaerospermum [24]*. In addition, antiviral activity was demonstrated as Enviro-Biotics® inactivated coronavirus virions (data not published). Spore dispersal is continuous and automatic, ensuring constant replication and reducing human intervention to a minimum. This extends the probiotic coverage to infrequently treated surfaces such as walls, fabrics, furniture, and basically the entire indoor environment. A continuous re-application of Bacillus supports a constant population of Enviro-Biotics® that treats surface organic contaminants and later acts as a protective layer against any newly introduced pathogens, preventing them from proliferating and colonizing the surface. The dispersal of the probiotics is carried out by localized stand-alone devices or incorporated into a building's HVAC system, thus serving as a holistic solution that treats both all the indoor spaces connected to the HVAC system as well as the HVAC system itself (ducts, filter, air outlets, etc.).

In parallel with anti-pathogen treatment, HS4U introduces dynamic control and extensive partitioning of the HVAC and sanitary systems. It is particularly relevant because during the COVID-19 crisis, there was evidence of contamination via the HVAC [9, 21] and the sewage systems [20, 22, 56], especially in high-rise buildings. In the post-COVID world, there have been extensive proposals for design optimization of HVAC [47, 48] and sewage systems [22] in order to prevent such transmissions in the future. HVAC optimization has also been proposed by Mihai and Rusu [35] in the specific context of ships.

Specifically, in the first phase, Hydrus plans to provide proof of concept that a small part of the ship – to be used as a confinement area – can be completely isolated by partitioning and automating the HVAC and sanitary systems. In a second phase, the possibility of extending this solution to the entire ship will be investigated. Such network partitioning aims to minimize contamination between passengers when a passenger is already contagious but not yet diagnosed and isolated. The work will also include economic considerations. Indeed, automation and optimal control of HVAC and sanitary systems can reduce energy and water usage and thus compensate for the added costs of automation. Let us emphasize that these solutions will be developed to be compatible with both new ships and ship retrofits.

For isolation measures to be effective, contact cases must be properly identified. Contact tracing and/or passenger localization will be necessary if a passenger is found to be ill, especially if their illness is contagious. Almeida et al. [2, this volume] review the options for achieving this on a cruise ship. This functionality is easily implemented for scenario 1 at no additional cost. For scenario 2 however, another sensor infrastructure may be required. Alternatively, passengers may be asked to activate contact tracking on their cell phones. Interestingly, most contact tracking applications operate without the exchange of personal data, which could increase the acceptability of contact tracking to passengers.

Finally, information management is critical in the event of an outbreak. The HS4U framework includes dashboards for crew and passengers as well as tools

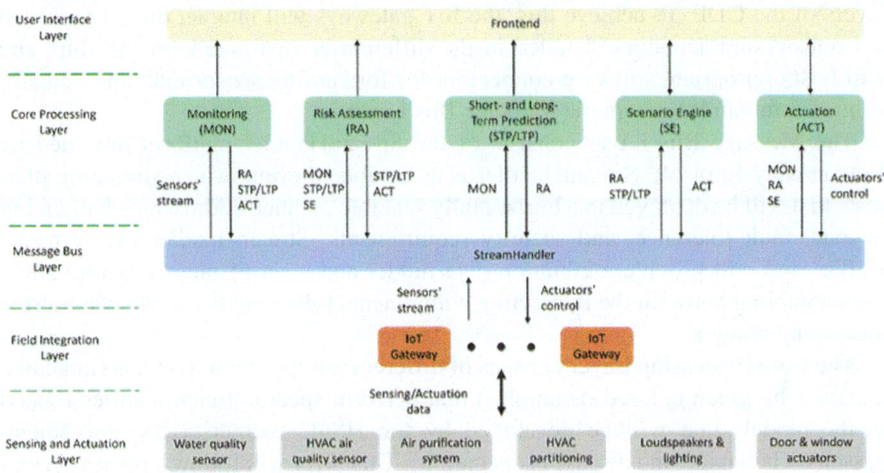

Fig. 7 High-level view of the interaction between S&A and the CDF components

for reporting to the relevant authorities. To enable crew or medical teams to easily communicate outbreak-related information to passengers, dashboards will be made available on mobile phones and on the various displays on board (in cabins and public areas). In regular operation (in the absence of an outbreak), they may also include personalized passenger information collected by the various sensors, similar to the dashboards proposed by activity trackers (e.g., activity level, health status, etc.). To streamline the crew efforts during an outbreak, crew dashboards available on tablets and smart phones will facilitate the transfer of information between different teams and advise the crew on actions to be taken.

5.4 *Digital Platform for S&A Data Processing*

To facilitate the use of S&A to prevent and mitigate outbreaks in an efficient, automated, and easily reconfigurable manner, HS4U will design and develop a Collaborative Digital Framework (CDF) that processes the sensor data and combines it with models of passenger behavior and disease propagation to suggest actions to be taken. Figure 7 presents a high-level view of the CDF, showing the different layers of functionality provided and the interactions between them.

The Sensing and Actuation Layer consists of all the S&A systems that will be either developed and/or upgraded in HS4U (e.g., pathogen detection systems, air purification systems, smart wearable devices) or already exist on ships (e.g., loudspeakers and other alarm systems, door/window actuators, etc.).

The Field Integration Layer consists of IoT gateways that provide the interoperability framework connecting the Sensing and Actuation Layer with the upper layers of the CDF. To achieve this, the IoT gateways will interact directly with all the sensors and actuators installed in the different compartments of the ship, and will host appropriate software connectors for forwarding sensor data and actuator control commands to/from the Message Bus Layer.

The Message Bus Layer consists of the StreamHandler platform provided by Netcompany-Intrasoft. StreamHandler is a distributed event-driven streaming platform that will be deployed in a horizontally scalable manner, addressing challenging in load, fault tolerance, and security requirements. StreamHandler uses Apache Kafka[5] and will provide a common data model and a set of interfaces that allow loose coupling between the interacting components following the publish/subscribe messaging pattern.

The Core Processing Layer consists of different groups of components (indicated in Fig. 7 by green colored rectangles) that perform specific functionalities related to the models that will be developed by the HS4U partners (risk assessment, passenger behavior, and disease propagation). This layer will follow a microservices architecture, where the various software components that make up the different sets of functionalities will be deployed in a containerized form (e.g., Docker), supported by a Kubernetes[6] instance for their management. The interactions between all these components will also take place via the Message Bus Layer.

Below is a brief description of the different component groups and their interactions.

Monitoring (MON) In MON, the platform receives input from all the sensing devices and harmonizes the observations into a readable form. The main service of the MON components is to monitor and measure the status of ship systems, pathogens in air, water and other locations, and the passenger conditions (e.g., fever). These measurements are then used by other components of the platform to trigger specific actions.

Risk Assessment (RA) In RA, measurements are categorized based on severity (risk assessment models, e.g., speed of disease spread, severity, passenger demographics). An ontology representation of the sensor measurements steers the whole process. If a high severity event is detected, an emergency scenario should be activated. RA models fall into two categories: a) strategic models for known pathogens, which address the entire ship, and b) tactical models for mitigation and response measures (e.g., what is the risk of quarantining passengers in one room instead of another). Essentially, the RA components receive thresholds and severity levels from the STP/LTP (see below) that guide the process to trigger actions and notifications to responsible crew members. In addition, prediction on the platform may not only be related to the emergency scenarios, but could also notify the crew of potentially dangerous situations (e.g., a healthy individual is about to enter an infested area).

[5] https://kafka.apache.org/.

[6] https://kubernetes.io.

Short and Long-term Prediction (STP/LTP) The STP/LTP is where the full lifecycle of the modeling process is implemented. Models for different objectives are executed in the form of containerized instances that provide valuable outputs to the other components. To implement the strategic and tactical risk assessment, the required predictive model includes pedestrian analytics, disease propagation, resource allocation, and other supporting models.

Scenario Engine (SE) Through the SE, the platform implements a user interface (UI) board that is capable of creating a schematic scenario indicating actionable "if-this-then-that" type of plans. These plans are created per ship by the crew and ship owners. In addition, scenarios/plans are reviewed in real time based on forecasts and tactical risk assessment in collaboration with STP/LTP and RA. The platform communicates the tactical risk assessment to the crew so that they can make more informed decisions. The decisions are co-created with the platform: the platform provides a simulation environment where the crew can virtually test decisions and actions ("what-ifs"), resulting in a new risk assessment.

Actuation (ACT) Through the ACT, the platform will assist crew members in carrying out a specific plan. This is done by automatically enabling available actuators on the ship, e.g., closing doors, switching off lights to darken rooms and deter passengers from entering, and activating ventilation.

Finally, the User Interface Layer constitutes the CDF's frontend to the different kinds of end users (e.g., captain, crew, passenger, system administrator). In particular, depending on the user's needs and access rights, the frontend will visualize various types of information coming from the components of the Core Processing Layer (e.g., through suitable menus, dashboards, and notifications) and also provide an interface for sending specific commands or requests.

6 Conclusion and Prospects

This chapter reports on the progress made by HS4U in defining a set of sensor and actuator solutions suitable for use on cruise ships to prevent and mitigate outbreaks. After presenting the high-level requirements of the solutions, including business and integration constraints, the chapter addresses the state of the art of sensors and actuators for outbreak management in and outside the naval context: technologies for symptom detection, disease recognition, and passenger localization are discussed, as well as approaches for surface, air, and duct purification, and for communication between the crew, the public, and authorities. Coupling the requirements with the state of the art allows to describe several feasible scenarios for the deployment of sensors and actuators aboard a cruise ship. The architecture of a digital platform capable of utilizing the proposed sensors and actuators to assist the crew in preventing and mitigating an outbreak is also described. While the work reported here has been essential to allow the consortium to enter the low-level requirements and design phase of its solutions, it may also have a broader impact: on the one hand, it aims to familiarize the cruise ship stakeholders with the technologies offered by the Internet of Things for health-related matters and to consider their adoption in the near future; on the other hand, it may contribute to the development of future regulations regarding outbreak management by competent authorities.

Acknowledgments This chapter acknowledges funding from European Union for the project HS4U under GA number 101069937 and from European Union and from Indian Government, Ministry of Science and Technology for the project LOTUS under Grant Agreement N° 820881. The authors thank Arturo Sommariva and Marco Berzieri from CNT Labs for their helpful inputs.

References

1. Al-Ghalith, G.A., Knights, D.: Bygiene: The new paradigm of bidirectional hygiene. Yale J. Biol. Med. **88**, 359–365 (2015)
2. Almeida, B., Mateus, M., Pacheco, M., Malo, P., Teixeira, T.: Technology approaches for cruise ship disease propagation monitoring. In: Rassia, S.T. (eds.) The Blue Book: Smart Sustainable Coastal Cities and Blue Growth Strategies for Marine and Maritime Environments, Springer, New York (this volume)
3. Aslam, S., Michaelides, M.P., Herodotou, H.: Internet of ships: A survey on architectures, emerging applications, and challenges. IEEE Internet Things J. **7**, 9714–9727 (2020)
4. Azara, A., Piana, A., Sotgiu, G., Dettori, M., Deriu, M.G., Masia, M.D., Are, B.M., Muresu, E.: Prevalence study of Legionella spp. contamination in ferries and cruise ships. BMC Public Health. **6**, 100 (2006)
5. Barsocchi, P., Ferro, E., La Rosa, D., Mahroo, A., Spoladore, D.: E-cabin: A software architecture for passenger comfort and cruise ship management. Sensors. **19**, 4978 (2019)
6. Caselli, E.: Hygiene: microbial strategies to reduce pathogens and drug resistance in clinical settings. Microb. Biotechnol. **10**, 1079–1083 (2017)
7. Citrawati, H., Barnabas, G.P.L., Santoso, T.A., Suharjito: Application internet of things in healthcare Covid-19: A systematic review. AIP Conference Proceedings 2594, 090012 (2023)
8. Contesini, F.J., de Melo, R.R., Sato, H.H.: An overview of Bacillus proteases: from production to application. Crit. Rev. Biotechnol. **38**, 321–334 (2018)
9. Correia, G., Rodrigues, L., Gameiro da Silva, M., Gonçalves, T.: Airborne route and bad use of ventilation systems as non-negligible factors in SARS-CoV-2 transmission. Med. Hypotheses. **141**, 109781 (2020)
10. Cousin, P., Moumtzidou, A., Karakostas, A., Gounaridis, L., Kouloumentas, C., Pereira, M.F., Apostolakis, A., Gorrochategui, P., Aoust, G., Lebental, B.: Improving water quality and security with advanced sensors and indirect water sensing methods. In: Di Mauro, A., Scozzari, A., Soldovieri, F. (eds.) Instrumentation and Measurement Technologies for Water Cycle Management, Springer Water, pp. 251–277. Springer International Publishing, Cham (2022)
11. D'Accolti, M., Soffritti, I., Bini, F., Mazziga, E., Cason, C., Comar, M., Volta, A., Bisi, M., Fumagalli, D., Mazzacane, S., Caselli, E.: Shaping the subway microbiome through probiotic-based sanitation during the COVID-19 emergency: a pre-post case-control study. Microbiome. **11**, 64 (2023)
12. D'Accolti, M., Soffritti, I., Piffanelli, M., Bisi, M., Mazzacane, S., Caselli, E.: Efficient removal of hospital pathogens from hard surfaces by a combined use of bacteriophages and probiotics: potential as sanitizing agents. Infect. Drug. Resist. **11**, 1015–1026 (2018)

13. Dbouk, T., Drikakis, D.: On coughing and airborne droplet transmission to humans. Phys. Fluids (1994) 32, 053310 (2020)
14. Dbouk, T., Drikakis, D.: Fluid dynamics and epidemiology: Seasonality and transmission dynamics. Phys. Fluids (1994) 33, 021901 (2021)
15. Dbouk, T., Roger, F., Drikakis, D.: Reducing indoor virus transmission using air purifiers. Phys. Fluids. **33**, 103301 (2021)
16. EU SHIPSAN Joint Act: European Manual for Hygiene Standards and Communicable Disease Surveillance on Passenger Ships [WWW Document]. EU SHIPSAN ACT Joint Action. URL https://www.shipsan.eu/Home/EuropeanManual.aspx (2016). Accessed 6.19.23
17. European Medicines Agency: Advice for cruise ship operators for preparedness and response to an outbreak of COVID-19 (2022)
18. Falagas, M.E., Makris, G.C.: Probiotic bacteria and biosurfactants for nosocomial infection control: a hypothesis. J. Hosp. Infect. **71**, 301–306 (2009)
19. Feng, Z., Wei, F., Yu, C.W., Li, H.: Valuation of indoor disinfection technologies for airborne disease control in hospital. Indoo. Built. Environ. **30**, 727–731 (2021)
20. Gormley, M., Aspray, T.J., Kelly, D.A.: COVID-19: mitigating transmission via wastewater plumbing systems. Lancet Glob. Health. **8**, e643 (2020)
21. Guo, M., Xu, P., Xiao, T., He, R., Dai, M., Miller, S.L.: Review and comparison of HVAC operation guidelines in different countries during the COVID-19 pandemic. Build. Environ. **187**, 107368 (2021)
22. Guo, Y., Li, X., Luby, S., Jiang, G.: Vertical outbreak of COVID-19 in high-rise buildings: The role of sewer stacks and prevention measures. Curr. Opin. Environ. Sci. Health. **29**, 100379 (2022)
23. Kabrein, H.A., Yusof, M.Z.M., Leman, A.M.: Progresses of filtration for removing particles and gases pollutants of indoor; limitations and future direction; review article. ARPN J. Eng. Appl. Sci. **11**, 7 (2016)
24. Kim, H.-S., Arellano, K., Park, H., Todorov, S.D., Kim, B., Kang, H., Park, Y.J., Suh, D.H., Jung, E.S., Ji, Y., Holzapfel, W.H.: Assessment of the safety and anti-inflammatory effects of three Bacillus strains in the respiratory tract. Environ. Microbiol. **23**, 3077–3098 (2021)
25. Kramer, A., Assadian, O.: Survival of microorganisms on inanimate surfaces. In: Borkow, G. (ed.) Use of Biocidal Surfaces for Reduction of Healthcare Acquired Infections, pp. 7–26. Springer International Publishing, Cham (2014)
26. Krokhine, S., Torabi, H., Doostmohammadi, A., Rezai, P.: Conventional and microfluidic methods for airborne virus isolation and detection. Colloids Surf. B: Biointerfaces. **206**, 111962 (2021)
27. Lebental, B.: Real-time, systematic disease detection on cruise ships: feasibility assessment for outbreak prevention. In: TBD. Springer Book (n.d.)
28. Liang, Y., Xiao, M., Xie, J., Li, J., Zhang, Yuyan, Liu, H., Zhang, Y., He, J., Zhang, G., Wei, N., Peng, L., Ke, Y., Zhang, Z.: Amplification-Free Detection of SARS-CoV-2 Down to Single Virus Level by Portable Carbon Nanotube Biosensors. Small **19**(34), 2208198 (2023). https://doi.org/10.1002/smll.202208198
29. Liston, P., Kay, A., Delemere, E., Karagiannidis, L., Kostovasili, M., Amditis, A., Drakoulis, D., Veltsistas, P.: Acceptability and usability of location-support technologies for cruise ship evacuations. In: Rauterberg, M., Fui-Hoon Nah, F., Siau, K., Krömker, H., Wei, J., Salvendy, G. (eds.) HCI International 2022 – Late Breaking Papers: HCI for Today's Community and Economy Lecture Notes in Computer Science, pp. 337–352. Springer Nature Switzerland, Cham (2022)
30. Liu, K., Yang, W., Chen, M., Zheng, K., Zeng, X., Zhang, S., Liu, C.: Deep-learning-based wireless human motion tracking for mobile ship environments. IEEE Internet Things J. **9**, 24186–24198 (2022)
31. Lv, Y., Hu, G., Wang, C., Yuan, W., Wei, S., Gao, J., Wang, B., Song, F.: Actual measurement, hygrothermal response experiment and growth prediction analysis of microbial contamination of central air conditioning system in Dalian, China. Sci. Rep. **7**, 44190 (2017)

32. Mah, T.-F., Pitts, B., Pellock, B., Walker, G.C., Stewart, P.S., O'Toole, G.A.: A genetic basis for Pseudomonas aeruginosa biofilm antibiotic resistance. Nature. **426**, 306–310 (2003)
33. Mannanov, R.N., Sattarova, R.K.: Antibiotics Produced by Bacillus Bacteria. Chem. Nat. Compd. **37**, 117–123 (2001)
34. Mendoza-Silva, G.M., Torres-Sospedra, J., Huerta, J.: A meta-review of indoor positioning systems. Sensors. **19**, 4507 (2019)
35. Mihai, V., Rusu, L.: An overview of the ship ventilation systems and measures to avoid the spread of diseases. Inventions. **6**, 55 (2021)
36. Muritala, B.A., Hernández-Lara, A.-B., Sánchez-Rebull, M.-V., Perera-Lluna, A.: #CoronavirusCruise: Impact and implications of the COVID-19 outbreaks on the perception of cruise tourism. Tour. Manag. Perspect. **41**, 100948 (2022)
37. Nolich, M., Spoladore, D., Carciotti, S., Buqi, R., Sacco, M.: Cabin as a home: A novel comfort optimization framework for IoT equipped smart environments and applications on cruise ships. Sensors. **19**, 1060 (2019)
38. Paik, B.-G., Cho, S.-R., Park, B.-J., Lee, D., Bae, B.-D.: Development of real-time monitoring system using wired and wireless networks in a full-scale ship. Engineering. **2**, 132–138 (2010)
39. Paik, B.-G., Cho, S.-R., Park, B.-J., Lee, D., Bae, B.-D., Yun, J.-H.: Characteristics of wireless sensor network for full-scale ship application. J. Mar. Sci. Technol. **14**, 115–126 (2009)
40. Pan, J., Hawks, S.A., Prussin, A.J., Duggal, N.K., Marr, L.C.: Collection, particle sizing and detection of airborne viruses. Environ. Sci. Technol. Lett. **9**, 71–76 (2022)
41. Pan, M., Lednicky, J.A., Wu, C.-Y.: Collection, particle sizing and detection of airborne viruses. J. Appl. Microbiol. **127**, 1596–1611 (2019)
42. Perrin, G., Lebental, B.: Uncertainty-based calibration method for environmental sensors—Application to chlorine and pH monitoring with carbon nanotube sensor array. IEEE Sensors J. **23**, 5146–5155 (2023)
43. Pinchuk, I.V., Bressollier, P., Verneuil, B., Fenet, B., Sorokulova, I.B., Mégraud, F., Urdaci, M.C.: In vitro anti-Helicobacter pylori activity of the probiotic strain Bacillus subtilis 3 is due to secretion of antibiotics. Antimicrob. Agents Chemother. **45**, 3156–3161 (2001)
44. Plaza-Hernandez, M., Sitton-Candanedo, I., Alonso, R.S., Martinez-de Iturrate, L.C., Prieto, J., Kravari, K., Kosmanis, T., Katranas, G., Silva, M.P., Corchado, J.M.: Edge computing and internet of things based platform to improve the quality of life of the silver economy on leisure cruise ships. In: 2021 International Symposium on Computer Science and Intelligent Controls (ISCSIC). Presented at the 2021 International Symposium on Computer Science and Intelligent Controls (ISCSIC), IEEE, Rome, Italy, pp. 159–163 (2021)
45. Romão, C., Miranda, C.A., Silva, J., Mandetta Clementino, M., de Filippis, I., Asensi, M.: Presence of qacEΔ1 Gene and Susceptibility to a Hospital Biocide in Clinical Isolates of Pseudomonas aeruginosa Resistant to Antibiotics. Curr. Microbiol. **63**, 16–21 (2011)
46. Sehulster, L., Chinn, R.Y.W.: Guidelines for environmental infection control in health-care facilities: recommendations of CDC and the Healthcare Infection Control Practices Advisory Committee (HICPAC) (No. 52(RR10);1–42). CDC and the Healthcare Infection Control Practices Advisory Committee (HICPAC) (2003)
47. Sha, H., Zhang, X., Qi, D.: Optimal control of high-rise building mechanical ventilation system for achieving low risk of COVID-19 transmission and ventilative cooling. Sustain. Cities Soc. **74**, 103256 (2021)
48. Shamim, J.A., Hsu, W.-L., Daiguji, H.: Review of component designs for post-COVID-19 HVAC systems: possibilities and challenges. Heliyon. **8**, e09001 (2022)
49. Sommariva, A., Cantarelli, B., Gagliardi, C.: Improving biological safety on ships using nanotechnology. In: Rassia, S.T. (eds.) The Blue Book: Smart Sustainable Coastal Cities and Blue Growth Strategies for Marine and Maritime Environments, Springer, New York (this volume)
50. Stone, W., Tolmay, J., Tucker, K., Wolfaardt, G.M.: Disinfectant, soap or probiotic cleaning? Surface microbiome diversity and biofilm competitive exclusion. Microorganisms. **8**, 1726 (2020)

51. Syriopoulos, T., Tsatsaronis, M., Gorila, M.: The global cruise industry: Financial performance evaluation. Res. Transp. Bus. Manag. **45**, 100558 (2022)
52. Tanikawa, T., Kiba, Y., Yu, J., Hsu, K., Chen, S., Ishii, A., Yokogawa, T., Suzuki, R., Inoue, Y., Kitamura, M.: Degradative effect of nattokinase on spike protein of SARS-CoV-2. Molecules. **27**, 5405 (2022)
53. Ursu, E.-L., Doroftei, F., Peptanariu, D., Pinteala, M., Rotaru, A.: DNA-assisted decoration of single-walled carbon nanotubes with gold nanoparticles for applications in surface-enhanced Raman scattering imaging of cells. J. Nanopart. Res. **19**, 181 (2017)
54. U.S. Department of Health and Human Services, U.S. Public Health Service, Centers for Disease Control and Prevention/National Center for Environmental Health: Vessel Sanitation Program 2018 Operations Manual (2018)
55. Vujovic, I., Kuzmanic, I., Kulenovic, Z.: Relationship of advances in electronics and maritime traffic, with case study of fall detection in smart cabins. Presented at the 7th International Maritime Science Conference, Solin, Croatia, pp. 193–199 (2017)
56. Wang, Q., Lin, Z., Niu, J., Choi, G.K.-Y., Fung, J.C.H., Lau, A.K.H., Louie, P., Leung, K.K.M., Huang, J., Cheng, P., Zhao, P., Chen, W., Zhang, S., Fu, L., Chan, P., Wong, A.H., Tse, H., Wong, S.C.Y., Lai, R.W.M., Hui, D.S., Yuen, K.-Y., Lung, D.C., Li, Y.: Spread of SARS-CoV-2 aerosols via two connected drainage stacks in a high-rise housing outbreak of COVID-19. J. Hazard. Mater. **430**, 128475 (2022)
57. Weber, D.J., Rutala, W.A., Miller, M.B., Huslage, K., Sickbert-Bennett, E.: Role of hospital surfaces in the transmission of emerging health care-associated pathogens: norovirus, Clostridium difficile, and Acinetobacter speciel. Am. J. Infect. Control. **38**, S25–S33 (2010)
58. World Health Organization: Infection prevention and control of epidemic-and pandemic prone acute respiratory infections in health care (2014)

Improving Biological Safety on Ships Using Nanotechnology

Arturo Sommariva, Bruno Cantarelli, and Costantino Gagliardi

1 Introduction

Nanotechnology is defined as the design, characterization, and application of structures, devices, and systems by controlling shape and size at a nanometer scale (1–100 nm) [1]. It is an emerging field of research, with various applications in science and technology, particularly for developing new materials. Nanoparticles are developed with unique properties that make them desirable in materials science and biology [1]. Among various nanoparticles, silver nanoparticles have been one of the most popular objects of study in recent decades [2]. Silver nanoparticles contain 20–15,000 silver atoms, and their diameters are usually smaller than 100 nm. Due to a large surface-to-volume ratio, silver nanoparticles exhibit remarkable antimicrobial activity, even at a low concentration [3]. In addition, they are low cost and have shown low cytotoxicity and immunological response [4]. Silver nanoparticles effect as an antimicrobial is strongly linked to their capability of emitting ions, since they have the proper reactivity to disturb or block fundamental bioprocesses that allow the microorganisms live and reproduce themselves. Silver nanoparticles are used instead of free ions due to the instability of free silver ions when exposed to light (visible spectrum and UVA, UVB, UVC), which make them precipitate as solid metallic silver (powder or mirror formation), inactivating the antimicrobial effect. The evolution of silver nanoparticles are the silver ion clusters: chemical objects stable in solution that are capable to stabilize silver ions when exposed to light and to the radiations cited before; in this case all the silver atoms are going to be effective and not only the ones that are freed from nanoparticles, this means that the concentration is easier to modulate and that a lower quantity of silver can be used to get the same effectiveness. This technology can be defined more as submicronic

A. Sommariva (✉) · B. Cantarelli · C. Gagliardi
CNT-Lab Srl, Italy
e-mail: consultant@cnt-lab.com; management@cnt-lab.com

S. Th. Rassia (ed.), *The Blue Book*, https://doi.org/10.1007/978-3-031-48831-3_12

than nano, because the dimension of these objects is from 100 nm to 1000 nm, their structure is dynamic, and changes in time as soon as it is kept together by interaction similar in energy to hydrogen bonds.

2 Antimicrobial Mechanism of Silver Ions

Although the exact mechanism of silver nanoparticles' antibacterial effects has not been entirely clarified. Owing to electrostatic attraction and affinity to sulfur proteins, silver ions can adhere to the cell wall and cytoplasmic membrane. The adhered ions can enhance the permeability of the cytoplasmic membrane and lead to disruption of the bacterial envelope [5]. After the uptake of free silver ions into cells, respiratory enzymes can be deactivated, generating reactive oxygen species but interrupting adenosine triphosphate production [6]. Reactive oxygen species can be a principal agent in the provocation of cell membrane disruption and deoxyribonucleic acid (DNA) modification. As sulfur and phosphorus are important components of DNA, the interaction of silver ions with the sulfur and phosphorus of DNA can cause problems in DNA replication, cell reproduction, or even result in termination of the microorganisms. Moreover, silver ions can inhibit the synthesis of proteins by denaturing ribosomes in the cytoplasm [7].The low cytotoxicity in animals and people results from the different nature of the cell membrane.

3 Silver Ions in Naval Sector

These materials first approached the naval sector to enhance the effectiveness of antifouling paints with the idea of producing antifouling paints without biocides. The main reasons were the high cost of the biocides due to the high concentration required to have the desired effect and the pollution generated from the degradation of those paints.

The first step of the research was the selection of the resin that resulted in choosing a one-component epoxy resin, solvent based.

The second step of the research consisted in testing the resistance of the final product (resin+silver clusters) in a special room saturated with salt spray at high temperature for 1 month.

The samples that performed better in the second step were tested in the sea for 6 months in La Spezia (Liguria) together with a sample of a classic biocides-based antifouling paint.

The results of this last test showed that one silver-based paint with a silver concentration $<0.1\%$ was much more effective than a standard biocides-based paint (Fig. 1).

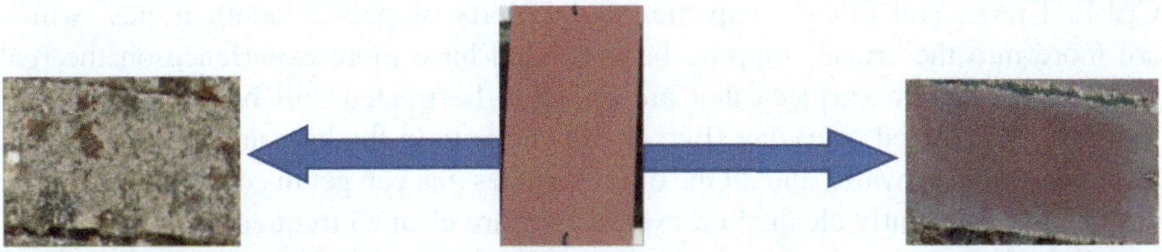

Fig. 1 Central image represent the sample as is. On the left is shown a sample coated with normal antifouling paint; on the right is shown a sample treated with silver clusters, both after 6 months inside sea water

4 A New Problem

After the beginning of the pandemic in 2020 with COVID-19, the interest shifted on the ship's inside as soon as there were big worries thinking about a lot of passengers, coming from different places and staying for long time in a relatively little space: the probability of contagion was considered really high.

Because of this, scientists and companies started to study new solutions to make environment as safer as possible.

After the against COVID-19 mass vaccination worries have subsided, but the pandemic changed the way in which people see the world, make them much more sensitive to the contagion problem of other diseases such influenza-like diseases, gastroenteritis (viral) and the more dangerous legionella and still COVID-19. Nowadays, people want to travel safe (in terms of contagion risk), but without taking the precautions they took during the pandemic.

In this scenario, the project and consortium HS4U (project 101069937) was born with the goal of making the cruise ship environment as safe as possible for the passenger, without distracting them from their vacation time.

5 The Role of CNT Lab in HS4U (Project 101069937)

CNT lab is an Italian company with more than 10 years of experience in silver-based technologies for biosafety of surfaces, with great results on different kind of plastics, paints, building materials, tissues, and so on. The technology used is the "silver ion cluster" synthAg™, patented by CNT lab in 2018, which, as explained above, is much more efficient than the classical nanoparticles and require a much less quantity of silver to get the same results, resulting in a much lower environmental impact.

The idea is to treat the most risky surfaces, that is, the surfaces that a large number of people can get in touch with, in a small range of time.

To know better where this risk is higher and on which surfaces concentrate our resources, CNT lab collaborated directly with LEDRA, COLUMBIA BLUE,

CELESTYAL, and other companies and experts of public health issues, which are more into the cruise shipping business and have more experience on the real problem. The main surfaces that are going to be treated will be carpets, tissues that are not changed everyday (like chairs and sofa in the bars and public places), switches, handles, walls, and all the other surfaces that can get in contact with people and are not frequently cleaned or, even if they are cleaned frequently, they get a lot of solicitations during the day, so that even cleaning them one time everyday would not be enough to guarantee their biological safety.

The surfaces are very different (tissue, plastic, metal) and require different techniques to treat them, also thinking about their position, their usage, and their aesthetics that must not be changed.

Tissues were treated with a “spray-and-dry” technique, which is a technique where a special liquid containing silver ion clusters, specially formulated for every tissue, is sprayed on the tissue and then dried in a ventilated oven, so that the liquid evaporate and silver clusters link with the molecules of the tissue, protecting them from the proliferation of microorganisms. The tests are performed with different concentration of silver, to find out which is the minimum efficient quantity to be sprayed to get the expected results.

In the case of the handles (for doors and windows mostly), the main material used is metal (stainless steel, aluminum, and brass) two possible solutions were identified: a “spray-and-dry” technique, with the difference that in this case the solvent will be more volatile as soon as they should be dried at room temperature, and a plastic transparent coating with a special material pretreated with silver ion clusters.

The switches are mostly made out of plastic and they are already fixed on the walls, so the main technique that was identified was the “transparent film coating” pretreated with silver ion clusters so that only the safe film will be touched by people, reducing the risk of hand-to-hand contagion.

Once all these surfaces are treated in these ways it is needed to make the cruise ship work regularly and collect results on the contagion to make a comparison with the previews without treatment data so that the efficacy of the treatments to preserve public health, without disturbing the vacation mood of the passengers, can be proved.

6 Conclusions

Laboratory tests show a good efficacy in terms of protection of the surfaces from the proliferation of microorganisms. The next step is to find the best way to put those treated materials in the real situation to be stress-tested in regular working conditions. After the analysis of the results, this technology application can be judged and, if possible, adjusted and optimized.

References

1. Kesharwani, P., Gorain, B., Low, S.Y., et al.: Nanotechnology based approaches for anti-diabetic drugs delivery. Diabetes Res. Clin. Pract. **136**, 52–77 (2018). https://doi.org/10.1016/j.diabres.2017.11.018
2. Saravanan, M., Barik, S.K., Mubarakali, D., Prakash, P., Pugazhendhi, A.: Synthesis of silver nanoparticles from Bacillus brevis (NCIM 2533) and their antibacterial activity against pathogenic bacteria. Microb. Pathog. **116**, 221–226 (2018). https://doi.org/10.1016/j.micpath.2018.01.038
3. Oves, M., Aslam, M., Rauf, M.A., et al.: Antimicrobial and anticancer activities of silver nanoparticles synthesized from the root hair extract of Phoenix dactylifera. Mater. Sci. Eng. C. **89**, 429–443 (2018). https://doi.org/10.1016/j.msec.2018.03.035
4. Samuel, M.S., Jose, S., Selvarajan, E., Mathimani, T., Pugazhendhi, A.: Biosynthesized silver nanoparticles using Bacillus amyloliquefaciens application for cytotoxicity effect on A549 cell line and photocatalytic degradation of p-nitrophenol. J. Photochem. Photobiol. **202**, 111642 (2020)
5. Khorrami, S., Zarrabi, A., Khaleghi, M., Danaei, M., Mozafari, M.: Selective cytotoxicity of green synthesized silver nanoparticles against the MCF-7 tumor cell line and their enhanced antioxidant and antimicrobial properties. Int. J. Nanomedicine. **13**, 8013–8024 (2018). https://doi.org/10.2147/IJN.S189295
6. Ramkumar, V.S., Pugazhendhi, A., Gopalakrishnan, K., et al.: Biofabrication and characterization of silver nanoparticles using aqueous extract of seaweed Enteromorpha compressa and its biomedical properties. Biotechnol. Rep. **14**, 1–7 (2017). https://doi.org/10.1016/j.btre.2017.02.001
7. Durán, N., Nakazato, G., Seabra, A.: Antimicrobial activity of biogenic silver nanoparticles, and silver chloride nanoparticles: an overview and comments. Appl. Microbiol. Biotechnol. **100**(15), 6555–6570 (2016). https://doi.org/10.1007/s00253-016-7657-7

The manufacturer's authorised representative in the EU is Springer Nature Customer Service Centre GmbH, Europaplatz 3, 69115 Heidelberg, Germany. If you have any concerns regarding our products, please contact ProductSafety@springernature.com

Printed and bound by CPI Group (UK) Ltd, Croydon, CR0 4YY
28/09/2026
02224075-0003